Springer Handbook
of Nanotechnology

Bharat Bhushan (Ed.)

3rd revised and extended edition

Springer Handbook of Nanotechnology

Since 2004 and with the 2nd edition in 2006, the Springer Handbook of Nanotechnology has established itself as the definitive reference in the nanoscience and nanotechnology area. It integrates the knowledge from nanofabrication, nanodevices, nanomechanics, nanotribology, materials science, and reliability engineering in just one volume. Beside the presentation of nanostructures, micro/nanofabrication, and micro/nanodevices, special emphasis is on scanning probe microscopy, nanotribology and nanomechanics, molecularly thick films, industrial applications and microdevice reliability, and on social aspects. In its 3rd edition, the book grew from 8 to 9 parts now including a part with chapters on biomimetics. More information is added to such fields as bionanotechnology, nanorobotics, and (bio) MEMS/NEMS, bio/nanotribology and bio/nanomechanics. The book is organized by an experienced editor with a universal knowledge and written by an international team of over 145 distinguished experts. It addresses mechanical and electrical engineers, materials scientists, physicists and chemists who work either in the nano area or in a field that is or will be influenced by this new key technology.

"The strong point is its focus on many of the practical aspects of nanotechnology... Anyone working in or learning about the field of nanotechnology would find this an excellent working handbook."

IEEE Electrical Insulation Magazine

"Outstandingly succeeds in its aim... It really is a magnificent volume and every scientific library and nanotechnology group should have a copy."

Materials World

"The integrity and authoritativeness... is guaranteed by an experienced editor and an international team of authors which have well summarized in their chapters information on fundamentals and applications."

Polymer News

List of Abbreviations
1 Introduction to Nanotechnology

Part A Nanostructures, Micro-/Nanofabrication and Materials
2 Nanomaterials Synthesis and Applications: Molecule-Based Devices
3 Introduction to Carbon Nanotubes
4 Nanowires
5 Template-Based Synthesis of Nanorod or Nanowire Arrays
6 Templated Self-Assembly of Particles
7 Three-Dimensional Nanostructure Fabrication by Focused Ion Beam Chemical Vapor Deposition
8 Introduction to Micro-/Nanofabrication
9 Nanoimprint Lithography-Patterning of Resists Using Molding
10 Stamping Techniques for Micro- and Nanofabrication
11 Material Aspects of Micro- and Nanoelectromechanical Systems

Part B MEMS/NEMS and BioMEMS/NEMS
12 MEMS/NEMS Devices and Applications
13 Next-Generation DNA Hybridization and Self-Assembly Nanofabrication Devices
14 Single-Walled Carbon Nanotube Sensor Concepts
15 Nanomechanical Cantilever Array Sensors
16 Biological Molecules in Therapeutic Nanodevices
17 G-Protein Coupled Receptors: Progress in Surface Display and Biosensor Technology
18 Microfluidic Devices and Their Applications to Lab-on-a-Chip
19 Centrifuge-Based Fluidic Platforms
20 Micro-/Nanodroplets in Microfluidic Devices

Part C Scanning-Probe Microscopy

21 Scanning Probe Microscopy-Principle of Operation, Instrumentation, and Probes
22 General and Special Probes in Scanning Microscopies
23 Noncontact Atomic Force Microscopy and Related Topics
24 Low-Temperature Scanning Probe Microscopy
25 Higher Harmonics and Time-Varying Forces in Dynamic Force Microscopy
26 Dynamic Modes of Atomic Force Microscopy
27 Molecular Recognition Force Microscopy: From Molecular Bonds to Complex Energy Landscapes

Part D Bio-/Nanotribology and Bio-/Nanomechanics

28 Nanotribology, Nanomechanics, and Materials Characterization
29 Surface Forces and Nanorheology of Molecularly Thin Films
30 Friction and Wear on the Atomic Scale
31 Computer Simulations of Nanometer-Scale Indentation and Friction
32 Force Measurements with Optical Tweezers
33 Scale Effect in Mechanical Properties and Tribology
34 Structural, Nanomechanical, and Nanotribological Characterization of Human Hair Using Atomic Force Microscopy and Nanoindentation
35 Cellular Nanomechanics
36 Optical Cell Manipulation
37 Mechanical Properties of Nanostructures

Part E Molecularly Thick Films for Lubrication

38 Nanotribology of Ultrathin and Hard Amorphous Carbon Films
39 Self-Assembled Monolayers for Nanotribology and Surface Protection
40 Nanoscale Boundary Lubrication Studies

Part F Biomimetics

41 Multifunctional Plant Surfaces and Smart Materials
42 Lotus Effect: Surfaces with Roughness-Induced Superhydrophobicity, Self-Cleaning, and Low Adhesion
43 Biological and Biologically Inspired Attachment Systems
44 Gecko Feet: Natural Hairy Attachment Systems for Smart Adhesion

Part G Industrial Applications

45 The *Millipede*-A Nanotechnology-Based AFM Data-Storage System
46 Nanorobotics

Part H Micro-/Nanodevice Reliability

47 MEMS/NEMS and BioMEMS/BioNEMS: Materials, Devices, and Biomimetics
48 Friction and Wear in Micro- and Nanomachines
49 Failure Mechanisms in MEMS/NEMS Devices
50 Mechanical Properties of Micromachined Structures
51 High-Volume Manufacturing and Field Stability of MEMS Products
52 Packaging and Reliability Issues in Micro-/Nanosystems

Part I Technological Convergence and Governing Nanotechnology

53 Governing Nanotechnology: Social, Ethical and Human Issues

Subject Index

使 用 说 明

1.《纳米技术手册》原版为一册，分为A~I部分。考虑到使用方便以及内容一致，影印版分为7册：第1册—Part A，第2册—Part B，第3册—Part C，第4册—Part D，第5册—Part E，第6册—Part F，第7册—Part G、H、I。

2.各册在页脚重新编排页码，该页码对应中文目录。保留了原书页眉及页码，其页码对应原书目录及主题索引。

3.各册均给出完整7册书的章目录。

4.作者及其联系方式、缩略语表各册均完整呈现。

5.主题索引安排在第7册。

6.目录等采用中英文对照形式给出，方便读者快速浏览。

材料科学与工程图书工作室

联系电话　0451-86412421
　　　　　0451-86414559
邮　　箱　yh_bj@yahoo.com.cn
　　　　　xuyaying81823@gmail.com
　　　　　zhxh6414559@yahoo.com.cn

Springer 手册精选系列

纳米技术手册

厚膜分子润滑

【第5册】

Springer
Handbook of
Nanotechnology

〔美〕Bharat Bhushan 主编

（第三版影印版）

哈尔滨工业大学出版社
HARBIN INSTITUTE OF TECHNOLOGY PRESS

黑版贸审字 08-2013-001号

Reprint from English language edition:
Springer Handbook of Nanotechnology
by Bharat Bhushan
Copyright © 2010 Springer Berlin Heidelberg
Springer Berlin Heidelberg is a part of Springer Science+Business Media
All Rights Reserved

This reprint has been authorized by Springer Science & Business Media for distribution in China Mainland only and not for export there from.

图书在版编目（CIP）数据

纳米技术手册：第3版. 5, 厚膜分子润滑 =Handbook of Nanotechnology. 5, Molecularly Thickfilms for Lubrication：英文 /（美）布尚(Bhushan,B.) 主编. —影印本. —哈尔滨：哈尔滨工业大学出版社, 2013.1
（Springer手册精选系列）
ISBN 978-7-5603-3951-1

Ⅰ.①纳… Ⅱ.①布… Ⅲ.①纳米技术－手册－英文②纳米润滑－手册－英文 Ⅳ.①TB303-62②TH117.2-62

中国版本图书馆CIP数据核字(2013)第004229号

责任编辑	杨　桦　许雅莹　张秀华	
出版发行	哈尔滨工业大学出版社	
社　　址	哈尔滨市南岗区复华四道街10号 邮编 150006	
传　　真	0451-86414749	
网　　址	http://hitpress.hit.edu.cn	
印　　刷	哈尔滨市石桥印务有限公司	
开　　本	787mm×960mm　1/16　印张 10.5	
版　　次	2013年1月第1版　2013年1月第1次印刷	
书　　号	ISBN 978-7-5603-3951-1	
定　　价	30.00元	

（如因印刷质量问题影响阅读，我社负责调换）

Foreword by Neal Lane

In a January 2000 speech at the California Institute of Technology, former President W.J. Clinton talked about the exciting promise of *nanotechnology* and the importance of expanding research in nanoscale science and engineering and, more broadly, in the physical sciences. Later that month, he announced in his State of the Union Address an ambitious US$ 497 million federal, multiagency national nanotechnology initiative (NNI) in the fiscal year 2001 budget; and he made the NNI a top science and technology priority within a budget that emphasized increased investment in US scientific research. With strong bipartisan support in Congress, most of this request was appropriated, and the NNI was born. Often, federal budget initiatives only last a year or so. It is most encouraging that the NNI has remained a high priority of the G.W. Bush Administration and Congress, reflecting enormous progress in the field and continued strong interest and support by industry.

Nanotechnology is the ability to manipulate individual atoms and molecules to produce nanostructured materials and submicron objects that have applications in the real world. Nanotechnology involves the production and application of physical, chemical and biological systems at scales ranging from individual atoms or molecules to about 100 nm, as well as the integration of the resulting nanostructures into larger systems. Nanotechnology is likely to have a profound impact on our economy and society in the early 21st century, perhaps comparable to that of information technology or cellular and molecular biology. Science and engineering research in nanotechnology promises breakthroughs in areas such as materials and manufacturing, electronics, medicine and healthcare, energy and the environment, biotechnology, information technology and national security. Clinical trials are already underway for nanomaterials that offer the promise of cures for certain cancers. It is widely felt that nanotechnology will be the next industrial revolution.

Nanometer-scale features are built up from their elemental constituents. Micro- and nanosystems components are fabricated using batch-processing techniques that are compatible with integrated circuits and range in size from micro- to nanometers. Micro- and nanosystems include micro/nanoelectro-mechanical systems (MEMS/NEMS), micromechatronics, optoelectronics, microfluidics and systems integration. These systems can sense, control, and activate on the micro/nanoscale and can function individually or in arrays to generate effects on the macroscale. Due to the enabling nature of these systems and the significant impact they can have on both the commercial and defense applications, industry as well as the federal government have taken special interest in seeing growth nurtured in this field. Micro- and nanosystems are the next logical step in the *silicon revolution*.

The discovery of novel materials, processes, and phenomena at the nanoscale and the development of new experimental and theoretical techniques for research provide fresh opportunities for the development of innovative nanosystems and nanostructured materials. There is an increasing need for a multidisciplinary, systems-oriented approach to manufacturing micro/nanodevices which function reliably. This can only be achieved through the cross-fertilization of ideas from different disciplines and the systematic flow of information and people among research groups.

Nanotechnology is a broad, highly interdisciplinary, and still evolving field. Covering even the most important aspects of nanotechnology in a single book that reaches readers ranging from students to active researchers in academia and industry is an enormous challenge. To prepare such a wide-ranging book on nanotechnology, Prof. Bhushan has harnessed his own knowledge and experience, gained in several industries and universities, and has assembled internationally recognized authorities from four continents to write chapters covering a wide array of nanotechnology topics, including the latest advances. The authors come from both academia and industry. The topics include major advances in many fields where nanoscale science and engineering is being pursued and illustrate how the field of nanotechnology has continued to emerge and blossom. Given the accelerating pace of discovery and applications in nanotechnology, it is a challenge to cap-

Prof. Neal Lane
Malcolm Gillis University Professor,
Department of Physics and Astronomy,
Senior Fellow,
James A. Baker III Institute for Public Policy
Rice University
Houston, Texas

Served in the Clinton Administration as Assistant to the President for Science and Technology and Director of the White House Office of Science and Technology Policy (1998–2001) and, prior to that, as Director of the National Science Foundation (1993–1998). While at the White House, he was a key figure in the creation of the NNI.

ture it all in one volume. As in earlier editions, professor Bhushan does an admirable job.

Professor Bharat Bhushan's comprehensive book is intended to serve both as a textbook for university courses as well as a reference for researchers. The first and second editions were timely additions to the literature on nanotechnology and stimulated further interest in this important new field, while serving as invaluable resources to members of the international scientific and industrial community. The increasing demand for up-to-date information on this fast moving field led to this third edition. It is increasingly important that scientists and engineers, whatever their specialty, have a solid grounding in the fundamentals and potential applications of nanotechnology. This third edition addresses that need by giving particular attention to the widening audience of readers. It also includes a discussion of the social, ethical and political issues that tend to surround any emerging technology.

The editor and his team are to be warmly congratulated for bringing together this exclusive, timely, and useful nanotechnology handbook.

Foreword by James R. Heath

Nanotechnology has become an increasingly popular buzzword over the past five years or so, a trend that has been fueled by a global set of publicly funded nanotechnology initiatives. Even as researchers have been struggling to demonstrate some of the most fundamental and simple aspects of this field, the term nanotechnology has entered into the public consciousness through articles in the popular press and popular fiction. As a consequence, the expectations of the public are high for nanotechnology, even while the actual public definition of nanotechnology remains a bit fuzzy.

Why shouldn't those expectations be high? The late 1990s witnessed a major information technology (IT) revolution and a minor biotechnology revolution. The IT revolution impacted virtually every aspect of life in the western world. I am sitting on an airplane at 30 000 feet at the moment, working on my laptop, as are about half of the other passengers on this plane. The plane itself is riddled with computational and communications equipment. As soon as we land, many of us will pull out cell phones, others will check e-mail via wireless modem, some will do both. This picture would be the same if I was landing in Los Angeles, Beijing, or Capetown. I will probably never actually print this text, but will instead submit it electronically. All of this was unthinkable a dozen years ago. It is therefore no wonder that the public expects marvelous things to happen quickly. However, the science that laid the groundwork for the IT revolution dates back 60 years or more, with its origins in fundamental solid-state physics.

By contrast, the biotech revolution was relatively minor and, at least to date, not particularly effective. The major diseases that plagued mankind a quarter century ago are still here. In some third-world countries, the average lifespan of individuals has actually decreased from where it was a full century ago. While the costs of electronics technologies have plummeted, health care costs have continued to rise. The biotech revolution may have a profound impact, but the task at hand is substantially more difficult than what was required for the IT revolution. In effect, the IT revolution was based on the advanced engineering of two-dimensional digital circuits constructed from relatively simple components – extended solids. The biotech revolution is really dependent upon the ability to reverse engineer three-dimensional analog systems constructed from quite complex components – proteins. Given that the basic science behind biotech is substantially younger than the science that has supported IT, it is perhaps not surprising that the biotech revolution has not really been a proper revolution yet, and it likely needs at least another decade or so to come into fruition.

Where does nanotechnology fit into this picture? In many ways, nanotechnology depends upon the ability to engineer two- and three-dimensional systems constructed from complex components such as macromolecules, biomolecules, nanostructured solids, etc. Furthermore, in terms of patents, publications, and other metrics that can be used to gauge the birth and evolution of a field, nanotech lags some 15–20 years behind biotech. Thus, now is the time that the fundamental science behind nanotechnology is being explored and developed. Nevertheless, progress with that science is moving forward at a dramatic pace. If the scientific community can keep up this pace and if the public sector will continue to support this science, then it is possible, and even perhaps likely, that in 20 years we may be speaking of the nanotech revolution.

The first edition of Springer Handbook of Nanotechnology was timely to assemble chapters in the broad field of nanotechnology. Given the fact that the second edition was in press one year after the publication of the first edition in April 2004, it is clear that the handbook has shown to be a valuable reference for experienced researchers as well as for a novice in the field. The third edition has one Part added and an expanded scope should have a wider appeal.

Prof. James R. Heath
Department of Chemistry
California Institute of Technology
Pasadena, California

Worked in the group of Nobel Laureate Richard E. Smalley at Rice University (1984–88) and co-invented Fullerene molecules which led to a revolution in Chemistry including the realization of nanotubes. The work on Fullerene molecules was cited for the 1996 Nobel Prize in Chemistry. Later he joined the University of California at Los Angeles (1994–2002), and co-founded and served as a Scientific Director of The California Nanosystems Institute.

Preface to the 3rd Edition

On December 29, 1959 at the California Institute of Technology, Nobel Laureate Richard P. Feynman gave at talk at the Annual meeting of the American Physical Society that has become one of the 20th century classic science lectures, titled *There's Plenty of Room at the Bottom*. He presented a technological vision of extreme miniaturization in 1959, several years before the word *chip* became part of the lexicon. He talked about the problem of manipulating and controlling things on a small scale. Extrapolating from known physical laws, Feynman envisioned a technology using the ultimate toolbox of nature, building nanoobjects atom by atom or molecule by molecule. Since the 1980s, many inventions and discoveries in fabrication of nanoobjects have been testament to his vision. In recognition of this reality, National Science and Technology Council (NSTC) of the White House created the Interagency Working Group on Nanoscience, Engineering and Technology (IWGN) in 1998. In a January 2000 speech at the same institute, former President W.J. Clinton talked about the exciting promise of *nanotechnology* and the importance of expanding research in nanoscale science and technology, more broadly. Later that month, he announced in his State of the Union Address an ambitious US$ 497 million federal, multi-agency national nanotechnology initiative (NNI) in the fiscal year 2001 budget, and made the NNI a top science and technology priority. The objective of this initiative was to form a broad-based coalition in which the academe, the private sector, and local, state, and federal governments work together to push the envelop of nanoscience and nanoengineering to reap nanotechnology's potential social and economic benefits.

The funding in the US has continued to increase. In January 2003, the US senate introduced a bill to establish a National Nanotechnology Program. On December 3, 2003, President George W. Bush signed into law the 21st Century Nanotechnology Research and Development Act. The legislation put into law programs and activities supported by the National Nanotechnology Initiative. The bill gave nanotechnology a permanent home in the federal government and authorized US$ 3.7 billion to be spent in the four year period beginning in October 2005, for nanotechnology initiatives at five federal agencies. The funds would provide grants to researchers, coordinate R&D across five federal agencies (National Science Foundation (NSF), Department of Energy (DOE), NASA, National Institute of Standards and Technology (NIST), and Environmental Protection Agency (EPA)), establish interdisciplinary research centers, and accelerate technology transfer into the private sector. In addition, Department of Defense (DOD), Homeland Security, Agriculture and Justice as well as the National Institutes of Health (NIH) also fund large R&D activities. They currently account for more than one-third of the federal budget for nanotechnology.

European Union (EU) made nanosciences and nanotechnologies a priority in Sixth Framework Program (FP6) in 2002 for a period of 2003–2006. They had dedicated small funds in FP4 and FP5 before. FP6 was tailored to help better structure European research and to cope with the strategic objectives set out in Lisbon in 2000. Japan identified nanotechnology as one of its main research priorities in 2001. The funding levels increases sharply from US$ 400 million in 2001 to around US$ 950 million in 2004. In 2003, South Korea embarked upon a ten-year program with around US$ 2 billion of public funding, and Taiwan has committed around US$ 600 million of public funding over six years. Singapore and China are also investing on a large scale. Russia is well funded as well.

Nanotechnology literally means any technology done on a nanoscale that has applications in the real world. Nanotechnology encompasses production and application of physical, chemical and biological systems at scales, ranging from individual atoms or molecules to submicron dimensions, as well as the integration of the resulting nanostructures into larger systems. Nanotechnology is likely to have a profound impact on our economy and society in the early 21st century, comparable to that of semiconductor technology, information technology, or cellular and molecular biology. Science and technology research in nanotechnology promises breakthroughs in areas such as materials and manufacturing, nanoelectronics, medicine and healthcare, energy, biotechnology, information technology and national security. It is widely felt that nanotechnology will be the next industrial revolution.

There is an increasing need for a multidisciplinary, system-oriented approach to design and manufacturing of micro/nanodevices which function reliably. This can only be achieved through the cross-fertilization of ideas from different disciplines and the systematic flow of information and people among research groups. Reliability is a critical technology for many micro- and nanosystems and nanostructured materials. A broad based handbook was needed, and the first edition of Springer Handbook of Nanotechnology was published in April 2004. It presented an overview of nanomaterial synthesis, micro/nanofabrication, micro- and nanocomponents and systems, scanning probe microscopy, reliability issues (including nanotribology and nanomechanics) for nanotechnology, and industrial applications. When the handbook went for sale in Europe, it was sold out in ten days. Reviews on the handbook were very flattering.

Given the explosive growth in nanoscience and nanotechnology, the publisher and the editor decided to develop a second edition after merely six months of publication of the first edition. The second edition (2007) came out in December 2006. The publisher and the editor again decided to develop a third edition after six month of publication of the second edition. This edition of the handbook integrates the knowledge from nanostructures, fabrication, materials science, devices, and reliability point of view. It covers various industrial applications. It also addresses social, ethical, and political issues. Given the significant interest in biomedical applications, and biomimetics a number of additional chapters in this arena have been added. The third edition consists of 53 chapters (new 10, revised 28, and as is 15). The chapters have been written by 139 internationally recognized experts in the field, from academia, national research labs, and industry, and from all over the world.

This handbook is intended for three types of readers: graduate students of nanotechnology, researchers in academia and industry who are active or intend to become active in this field, and practicing engineers and scientists who have encountered a problem and hope to solve it as expeditiously as possible. The handbook should serve as an excellent text for one or two semester graduate courses in nanotechnology in mechanical engineering, materials science, applied physics, or applied chemistry.

We embarked on the development of third edition in June 2007, and we worked very hard to get all the chapters to the publisher in a record time of about 12 months. I wish to sincerely thank the authors for offering to write comprehensive chapters on a tight schedule. This is generally an added responsibility in the hectic work schedules of researchers today. I depended on a large number of reviewers who provided critical reviews. I would like to thank Dr. Phillip J. Bond, Chief of Staff and Under Secretary for Technology, US Department of Commerce, Washington, D.C. for suggestions for chapters as well as authors in the handbook. Last but not the least, I would like to thank my secretary Caterina Runyon-Spears for various administrative duties and her tireless efforts are highly appreciated.

I hope that this handbook will stimulate further interest in this important new field, and the readers of this handbook will find it useful.

February 2010 Bharat Bhushan
Editor

Preface to the 2nd Edition

On 29 December 1959 at the California Institute of Technology, Nobel Laureate Richard P. Feynman gave at talk at the Annual meeting of the American Physical Society that has become one of the 20th century classic science lectures, titled "There's Plenty of Room at the Bottom." He presented a technological vision of extreme miniaturization in 1959, several years before the word "chip" became part of the lexicon. He talked about the problem of manipulating and controlling things on a small scale. Extrapolating from known physical laws, Feynman envisioned a technology using the ultimate toolbox of nature, building nanoobjects atom by atom or molecule by molecule. Since the 1980s, many inventions and discoveries in the fabrication of nanoobjects have been a testament to his vision. In recognition of this reality, the National Science and Technology Council (NSTC) of the White House created the Interagency Working Group on Nanoscience, Engineering and Technology (IWGN) in 1998. In a January 2000 speech at the same institute, former President W. J. Clinton talked about the exciting promise of "nanotechnology" and the importance of expanding research in nanoscale science and, more broadly, technology. Later that month, he announced in his State of the Union Address an ambitious $497 million federal, multiagency national nanotechnology initiative (NNI) in the fiscal year 2001 budget, and made the NNI a top science and technology priority. The objective of this initiative was to form a broad-based coalition in which the academe, the private sector, and local, state, and federal governments work together to push the envelope of nanoscience and nanoengineering to reap nanotechnology's potential social and economic benefits.

The funding in the U.S. has continued to increase. In January 2003, the U. S. senate introduced a bill to establish a National Nanotechnology Program. On 3 December 2003, President George W. Bush signed into law the 21st Century Nanotechnology Research and Development Act. The legislation put into law programs and activities supported by the National Nanotechnology Initiative. The bill gave nanotechnology a permanent home in the federal government and authorized $3.7 billion to be spent in the four year period beginning in October 2005, for nanotechnology initiatives at five federal agencies. The funds would provide grants to researchers, coordinate R&D across five federal agencies (National Science Foundation (NSF), Department of Energy (DOE), NASA, National Institute of Standards and Technology (NIST), and Environmental Protection Agency (EPA)), establish interdisciplinary research centers, and accelerate technology transfer into the private sector. In addition, Department of Defense (DOD), Homeland Security, Agriculture and Justice as well as the National Institutes of Health (NIH) would also fund large R&D activities. They currently account for more than one-third of the federal budget for nanotechnology.

The European Union made nanosciences and nanotechnologies a priority in the Sixth Framework Program (FP6) in 2002 for the period of 2003-2006. They had dedicated small funds in FP4 and FP5 before. FP6 was tailored to help better structure European research and to cope with the strategic objectives set out in Lisbon in 2000. Japan identified nanotechnology as one of its main research priorities in 2001. The funding levels increased sharply from $400 million in 2001 to around $950 million in 2004. In 2003, South Korea embarked upon a ten-year program with around $2 billion of public funding, and Taiwan has committed around $600 million of public funding over six years. Singapore and China are also investing on a large scale. Russia is well funded as well.

Nanotechnology literally means any technology done on a nanoscale that has applications in the real world. Nanotechnology encompasses production and application of physical, chemical and biological systems at scales, ranging from individual atoms or molecules to submicron dimensions, as well as the integration of the resulting nanostructures into larger systems. Nanotechnology is likely to have a profound impact on our economy and society in the early 21st century, comparable to that of semiconductor technology, information technology, or cellular and molecular biology. Science and technology research in nanotechnology promises breakthroughs in areas such as materials and manufacturing, nanoelectronics, medicine and healthcare, energy, biotechnology, information technology and national security. It is widely felt that nanotechnology will be the next industrial revolution.

There is an increasing need for a multidisciplinary, system-oriented approach to design and manufacturing of micro/nanodevices which function reliably. This can only be achieved through the cross-fertilization of ideas from different disciplines and the systematic flow of information and people among research groups. Reliability is a critical technology for many micro- and nanosystems and nanostructured materials. A broad based handbook was needed, and the first edition of Springer Handbook of Nanotechnology was published in April 2004. It presented an overview of nanomaterial synthesis, micro/nanofabrication, micro- and nanocomponents and systems, scanning probe microscopy, reliability issues (including nanotribology and nanomechanics) for nanotechnology, and industrial applications. When the handbook went for sale in Europe, it was sold out in ten days. Reviews on the handbook were very flattering.

Given the explosive growth in nanoscience and nanotechnology, the publisher and the editor decided to develop a second edition after merely six months of publication of the first edition. The second edition (2007) came out in December 2006. The publisher and the editor again decided to develop a third edition after six month of publication of the second edition. This edition of the handbook integrates the knowledge from nanostructures, fabrication, materials science, devices, and reliability point of view. It covers various industrial applications. It also addresses social, ethical, and political issues. Given the significant interest in biomedical applications, and biomimetics a number of additional chapters in this arena have been added. The third edition consists of 53 chapters (new 10, revised 28, and as is 15). The chapters have been written by 139 internationally recognized experts in the field, from academia, national research labs, and industry, and from all over the world.

This handbook is intended for three types of readers: graduate students of nanotechnology, researchers in academia and industry who are active or intend to become active in this field, and practicing engineers and scientists who have encountered a problem and hope to solve it as expeditiously as possible. The handbook should serve as an excellent text for one or two semester graduate courses in nanotechnology in mechanical engineering, materials science, applied physics, or applied chemistry.

We embarked on the development of third edition in June 2007, and we worked very hard to get all the chapters to the publisher in a record time of about 12 months. I wish to sincerely thank the authors for offering to write comprehensive chapters on a tight schedule. This is generally an added responsibility in the hectic work schedules of researchers today. I depended on a large number of reviewers who provided critical reviews. I would like to thank Dr. Phillip J. Bond, Chief of Staff and Under Secretary for Technology, US Department of Commerce, Washington, D.C. for suggestions for chapters as well as authors in the handbook. Last but not the least, I would like to thank my secretary Caterina Runyon-Spears for various administrative duties and her tireless efforts are highly appreciated.

I hope that this handbook will stimulate further interest in this important new field, and the readers of this handbook will find it useful.

February 2010   Bharat Bhushan
                                                                  Editor

Preface to the 1st Edition

On December 29, 1959 at the California Institute of Technology, Nobel Laureate Richard P. Feynman gave a talk at the Annual meeting of the American Physical Society that has become one classic science lecture of the 20th century, titled "There's Plenty of Room at the Bottom." He presented a technological vision of extreme miniaturization in 1959, several years before the word "chip" became part of the lexicon. He talked about the problem of manipulating and controlling things on a small scale. Extrapolating from known physical laws, Feynman envisioned a technology using the ultimate toolbox of nature, building nanoobjects atom by atom or molecule by molecule. Since the 1980s, many inventions and discoveries in fabrication of nanoobjects have been a testament to his vision. In recognition of this reality, in a January 2000 speech at the same institute, former President W. J. Clinton talked about the exciting promise of "nanotechnology" and the importance of expanding research in nanoscale science and engineering. Later that month, he announced in his State of the Union Address an ambitious $ 497 million federal, multi-agency national nanotechnology initiative (NNI) in the fiscal year 2001 budget, and made the NNI a top science and technology priority. Nanotechnology literally means any technology done on a nanoscale that has applications in the real world. Nanotechnology encompasses production and application of physical, chemical and biological systems at size scales, ranging from individual atoms or molecules to submicron dimensions as well as the integration of the resulting nanostructures into larger systems. Nanofabrication methods include the manipulation or self-assembly of individual atoms, molecules, or molecular structures to produce nanostructured materials and sub-micron devices. Micro- and nanosystems components are fabricated using top-down lithographic and nonlithographic fabrication techniques. Nanotechnology will have a profound impact on our economy and society in the early 21st century, comparable to that of semiconductor technology, information technology, or advances in cellular and molecular biology. The research and development in nanotechnology will lead to potential breakthroughs in areas such as materials and manufacturing, nanoelectronics, medicine and healthcare, energy, biotechnology, information technology and national security. It is widely felt that nanotechnology will lead to the next industrial revolution.

Reliability is a critical technology for many micro- and nanosystems and nanostructured materials. No book exists on this emerging field. A broad based handbook is needed. The purpose of this handbook is to present an overview of nanomaterial synthesis, micro/nanofabrication, micro- and nanocomponents and systems, reliability issues (including nanotribology and nanomechanics) for nanotechnology, and industrial applications. The chapters have been written by internationally recognized experts in the field, from academia, national research labs and industry from all over the world.

The handbook integrates knowledge from the fabrication, mechanics, materials science and reliability points of view. This book is intended for three types of readers: graduate students of nanotechnology, researchers in academia and industry who are active or intend to become active in this field, and practicing engineers and scientists who have encountered a problem and hope to solve it as expeditiously as possible. The handbook should serve as an excellent text for one or two semester graduate courses in nanotechnology in mechanical engineering, materials science, applied physics, or applied chemistry.

We embarked on this project in February 2002, and we worked very hard to get all the chapters to the publisher in a record time of about 1 year. I wish to sincerely thank the authors for offering to write comprehensive chapters on a tight schedule. This is generally an added responsibility in the hectic work schedules of researchers today. I depended on a large number of reviewers who provided critical reviews. I would like to thank Dr. Phillip J. Bond, Chief of Staff and Under Secretary for Technology, US Department of Commerce, Washington, D.C. for suggestions for chapters as well as authors in the handbook. I would also like to thank my colleague, Dr. Huiwen Liu, whose efforts during the preparation of this handbook were very useful.

I hope that this handbook will stimulate further interest in this important new field, and the readers of this handbook will find it useful.

September 2003   Bharat Bhushan
                                                                                             Editor

Editors Vita

Dr. Bharat Bhushan received an M.S. in mechanical engineering from the Massachusetts Institute of Technology in 1971, an M.S. in mechanics and a Ph.D. in mechanical engineering from the University of Colorado at Boulder in 1973 and 1976, respectively, an MBA from Rensselaer Polytechnic Institute at Troy, NY in 1980, Doctor Technicae from the University of Trondheim at Trondheim, Norway in 1990, a Doctor of Technical Sciences from the Warsaw University of Technology at Warsaw, Poland in 1996, and Doctor Honouris Causa from the National Academy of Sciences at Gomel, Belarus in 2000. He is a registered professional engineer. He is presently an Ohio Eminent Scholar and The Howard D. Winbigler Professor in the College of Engineering, and the Director of the Nanoprobe Laboratory for Bio- and Nanotechnology and Biomimetics (NLB²) at the Ohio State University, Columbus, Ohio. His research interests include fundamental studies with a focus on scanning probe techniques in the interdisciplinary areas of bio/nanotribology, bio/nanomechanics and bio/nanomaterials characterization, and applications to bio/nanotechnology and biomimetics. He is an internationally recognized expert of bio/nanotribology and bio/nanomechanics using scanning probe microscopy, and is one of the most prolific authors. He is considered by some a pioneer of the tribology and mechanics of magnetic storage devices. He has authored 6 scientific books, more than 90 handbook chapters, more than 700 scientific papers (h factor – 45+; ISI Highly Cited in Materials Science, since 2007), and more than 60 technical reports, edited more than 45 books, and holds 17 US and foreign patents. He is co-editor of Springer NanoScience and Technology Series and co-editor of Microsystem Technologies. He has given more than 400 invited presentations on six continents and more than 140 keynote/plenary addresses at major international conferences.

Dr. Bhushan is an accomplished organizer. He organized the first symposium on Tribology and Mechanics of Magnetic Storage Systems in 1984 and the first international symposium on Advances in Information Storage Systems in 1990, both of which are now held annually. He is the founder of an ASME Information Storage and Processing Systems Division founded in 1993 and served as the founding chair during 1993–1998. His biography has been listed in over two dozen Who's Who books including Who's Who in the World and has received more than two dozen awards for his contributions to science and technology from professional societies, industry, and US government agencies. He is also the recipient of various international fellowships including the Alexander von Humboldt Research Prize for Senior Scientists, Max Planck Foundation Research Award for Outstanding Foreign Scientists, and the Fulbright Senior Scholar Award. He is a foreign member of the International Academy of Engineering (Russia), Byelorussian Academy of Engineering and Technology and the Academy of Triboengineering of Ukraine, an honorary member of the Society of Tribologists of Belarus, a fellow of ASME, IEEE, STLE, and the New York Academy of Sciences, and a member of ASEE, Sigma Xi and Tau Beta Pi.

Dr. Bhushan has previously worked for the R&D Division of Mechanical Technology Inc., Latham, NY; the Technology Services Division of SKF Industries Inc., King of Prussia, PA; the General Products Division Laboratory of IBM Corporation, Tucson, AZ; and the Almaden Research Center of IBM Corporation, San Jose, CA. He has held visiting professor appointments at University of California at Berkeley, University of Cambridge, UK, Technical University Vienna, Austria, University of Paris, Orsay, ETH Zurich and EPFL Lausanne.

List of Authors

Chong H. Ahn
University of Cincinnati
Department of Electrical
and Computer Engineering
Cincinnati, OH 45221, USA
e-mail: *chong.ahn@uc.edu*

Boris Anczykowski
nanoAnalytics GmbH
Münster, Germany
e-mail: *anczykowski@nanoanalytics.com*

W. Robert Ashurst
Auburn University
Department of Chemical Engineering
Auburn, AL 36849, USA
e-mail: *ashurst@auburn.edu*

Massood Z. Atashbar
Western Michigan University
Department of Electrical
and Computer Engineering
Kalamazoo, MI 49008-5329, USA
e-mail: *massood.atashbar@wmich.edu*

Wolfgang Bacsa
University of Toulouse III (Paul Sabatier)
Laboratoire de Physique des Solides (LPST),
UMR 5477 CNRS
Toulouse, France
e-mail: *bacsa@ramansco.ups-tlse.fr;*
bacsa@lpst.ups-tlse.fr

Kelly Bailey
University of Adelaide
CSIRO Human Nutrition
Adelaide SA 5005, Australia
e-mail: *kelly.bailey@csiro.au*

William Sims Bainbridge
National Science Foundation
Division of Information, Science and Engineering
Arlington, VA, USA
e-mail: *wsbainbridge@yahoo.com*

Antonio Baldi
Institut de Microelectronica de Barcelona (IMB)
Centro National Microelectrónica (CNM-CSIC)
Barcelona, Spain
e-mail: *antoni.baldi@cnm.es*

Wilhelm Barthlott
University of Bonn
Nees Institute for Biodiversity of Plants
Meckenheimer Allee 170
53115 Bonn, Germany
e-mail: *barthlott@uni-bonn.de*

Roland Bennewitz
INM – Leibniz Institute for New Materials
66123 Saarbrücken, Germany
e-mail: *roland.bennewitz@inm-gmbh.de*

Bharat Bhushan
Ohio State University
Nanoprobe Laboratory for Bio- and
Nanotechnology and Biomimetics (NLB²)
201 W. 19th Avenue
Columbus, OH 43210-1142, USA
e-mail: *bhushan.2@osu.edu*

Gerd K. Binnig
Definiens AG
Trappentreustr. 1
80339 Munich, Germany
e-mail: *gbinnig@definiens.com*

Marcie R. Black
Bandgap Engineering Inc.
1344 Main St.
Waltham, MA 02451, USA
e-mail: *marcie@alum.mit.edu;*
marcie@bandgap.com

Donald W. Brenner
Department of Materials Science and Engineering
Raleigh, NC, USA
e-mail: *brenner@ncsu.edu*

Jean-Marc Broto
Institut National des Sciences Appliquées
of Toulouse
Laboratoire National
des Champs Magnétiques Pulsés (LNCMP)
Toulouse, France
e-mail: *broto@lncmp.fr*

Guozhong Cao
University of Washington
Dept. of Materials Science and Engineering
302M Roberts Hall
Seattle, WA 98195-2120, USA
e-mail: *gzcao@u.washington.edu*

Edin (I-Chen) Chen
National Central University
Institute of Materials Science and Engineering
Department of Mechanical Engineering
Chung-Li, 320, Taiwan
e-mail: *ichen@ncu.edu.tw*

Yu-Ting Cheng
National Chiao Tung University
Department of Electronics Engineering
& Institute of Electronics
1001, Ta-Hsueh Rd.
Hsinchu, 300, Taiwan, R.O.C.
e-mail: *ytcheng@mail.nctu.edu.tw*

Giovanni Cherubini
IBM Zurich Research Laboratory
Tape Technologies
8803 Rüschlikon, Switzerland
e-mail: *cbi@zurich.ibm.com*

Mu Chiao
Department of Mechanical Engineering
6250 Applied Science Lane
Vancouver, BC V6T 1Z4, Canada
e-mail: *muchiao@mech.ubc.ca*

Jin-Woo Choi
Louisiana State University
Department of Electrical
and Computer Engineering
Baton Rouge, LA 70803, USA
e-mail: *choi@ece.lsu.edu*

Tamara H. Cooper
University of Adelaide
CSIRO Human Nutrition
Adelaide SA 5005, Australia
e-mail: *tamara.cooper@csiro.au*

Alex D. Corwin
GE Global Research
1 Research Circle
Niskayuna, NY 12309, USA
e-mail: *corwin@ge.com*

Maarten P. de Boer
Carnegie Mellon University
Department of Mechanical Engineering
5000 Forbes Avenue
Pittsburgh, PA 15213, USA
e-mail: *mpdebo@andrew.cmu.edu*

Dietrich Dehlinger
Lawrence Livermore National Laboratory
Engineering
Livermore, CA 94551, USA
e-mail: *dehlinger1@llnl.gov*

Frank W. DelRio
National Institute of Standards and Technology
100 Bureau Drive, Stop 8520
Gaithersburg, MD 20899-8520, USA
e-mail: *frank.delrio@nist.gov*

Michel Despont
IBM Zurich Research Laboratory
Micro- and Nanofabrication
8803 Rüschlikon, Switzerland
e-mail: *dpt@zurich.ibm.com*

Lixin Dong
Michigan State University
Electrical and Computer Engineering
2120 Engineering Building
East Lansing, MI 48824-1226, USA
e-mail: *ldong@egr.msu.edu*

Gene Dresselhaus
Massachusetts Institute of Technology
Francis Bitter Magnet Laboratory
Cambridge, MA 02139, USA
e-mail: *gene@mgm.mit.edu*

Mildred S. Dresselhaus
Massachusetts Institute of Technology
Department of Electrical Engineering
and Computer Science
Department of Physics
Cambridge, MA, USA
e-mail: *millie@mgm.mit.edu*

Urs T. Dürig
IBM Zurich Research Laboratory
Micro-/Nanofabrication
8803 Rüschlikon, Switzerland
e-mail: *drg@zurich.ibm.com*

Andreas Ebner
Johannes Kepler University Linz
Institute for Biophysics
Altenberger Str. 69
4040 Linz, Austria
e-mail: *andreas.ebner@jku.at*

Evangelos Eleftheriou
IBM Zurich Research Laboratory
8803 Rüschlikon, Switzerland
e-mail: *ele@zurich.ibm.com*

Emmanuel Flahaut
Université Paul Sabatier
CIRIMAT, Centre Interuniversitaire de Recherche
et d'Ingénierie des Matériaux, UMR 5085 CNRS
118 Route de Narbonne
31062 Toulouse, France
e-mail: *flahaut@chimie.ups-tlse.fr*

Anatol Fritsch
University of Leipzig
Institute of Experimental Physics I
Division of Soft Matter Physics
Linnéstr. 5
04103 Leipzig, Germany
e-mail: *anatol.fritsch@uni-leipzig.de*

Harald Fuchs
Universität Münster
Physikalisches Institut
Münster, Germany
e-mail: *fuchsh@uni-muenster.de*

Christoph Gerber
University of Basel
Institute of Physics
National Competence Center for Research
in Nanoscale Science (NCCR) Basel
Klingelbergstr. 82
4056 Basel, Switzerland
e-mail: *christoph.gerber@unibas.ch*

Franz J. Giessibl
Universität Regensburg
Institute of Experimental and Applied Physics
Universitätsstr. 31
93053 Regensburg, Germany
e-mail: *franz.giessibl@physik.uni-regensburg.de*

Enrico Gnecco
University of Basel
National Center of Competence in Research
Department of Physics
Klingelbergstr. 82
4056 Basel, Switzerland
e-mail: *enrico.gnecco@unibas.ch*

Stanislav N. Gorb
Max Planck Institut für Metallforschung
Evolutionary Biomaterials Group
Heisenbergstr. 3
70569 Stuttgart, Germany
e-mail: *s.gorb@mf.mpg.de*

Hermann Gruber
University of Linz
Institute of Biophysics
Altenberger Str. 69
4040 Linz, Austria
e-mail: *hermann.gruber@jku.at*

Jason Hafner
Rice University
Department of Physics and Astronomy
Houston, TX 77251, USA
e-mail: *hafner@rice.edu*

Judith A. Harrison
U.S. Naval Academy
Chemistry Department
572 Holloway Road
Annapolis, MD 21402-5026, USA
e-mail: *jah@usna.edu*

Martin Hegner
CRANN – The Naughton Institute
Trinity College, University of Dublin
School of Physics
Dublin, 2, Ireland
e-mail: *martin.hegner@tcd.ie*

Thomas Helbling
ETH Zurich
Micro and Nanosystems
Department of Mechanical
and Process Engineering
8092 Zurich, Switzerland
e-mail: *thomas.helbling@micro.mavt.ethz.ch*

Michael J. Heller
University of California San Diego
Department of Bioengineering
Dept. of Electrical and Computer Engineering
La Jolla, CA, USA
e-mail: *mjheller@ucsd.edu*

Seong-Jun Heo
Lam Research Corp.
4650 Cushing Parkway
Fremont, CA 94538, USA
e-mail: *seongjun.heo@lamrc.com*

Christofer Hierold
ETH Zurich
Micro and Nanosystems
Department of Mechanical
and Process Engineering
8092 Zurich, Switzerland
e-mail: *christofer.hierold@micro.mavt.ethz.ch*

Peter Hinterdorfer
University of Linz
Institute for Biophysics
Altenberger Str. 69
4040 Linz, Austria
e-mail: *peter.hinterdorfer@jku.at*

Dalibor Hodko
Nanogen, Inc.
10498 Pacific Center Court
San Diego, CA 92121, USA
e-mail: *dhodko@nanogen.com*

Hendrik Hölscher
Forschungszentrum Karlsruhe
Institute of Microstructure Technology
Linnéstr. 5
76021 Karlsruhe, Germany
e-mail: *hendrik.hoelscher@imt.fzk.de*

Hirotaka Hosoi
Hokkaido University
Creative Research Initiative Sousei
Kita 21, Nishi 10, Kita-ku
Sapporo, Japan
e-mail: *hosoi@cris.hokudai.ac.jp*

Katrin Hübner
Staatliche Fachoberschule Neu-Ulm
89231 Neu-Ulm, Germany
e-mail: *katrin.huebner1@web.de*

Douglas L. Irving
North Carolina State University
Materials Science and Engineering
Raleigh, NC 27695-7907, USA
e-mail: *doug_irving@ncsu.edu*

Jacob N. Israelachvili
University of California
Department of Chemical Engineering
and Materials Department
Santa Barbara, CA 93106-5080, USA
e-mail: *jacob@engineering.ucsb.edu*

Guangyao Jia
University of California, Irvine
Department of Mechanical
and Aerospace Engineering
Irvine, CA, USA
e-mail: *gjia@uci.edu*

Sungho Jin
University of California, San Diego
Department of Mechanical
and Aerospace Engineering
9500 Gilman Drive
La Jolla, CA 92093-0411, USA
e-mail: *jin@ucsd.edu*

Anne Jourdain
Interuniversity Microelectronics Center (IMEC)
Leuven, Belgium
e-mail: *jourdain@imec.be*

Yong Chae Jung
Samsung Electronics C., Ltd.
Senior Engineer Process Development Team
San #16 Banwol-Dong, Hwasung-City
Gyeonggi-Do 445-701, Korea
e-mail: *yc423.jung@samsung.com*

Harold Kahn
Case Western Reserve University
Department of Materials Science and Engineering
Cleveland, OH, USA
e-mail: *kahn@cwru.edu*

Roger Kamm
Massachusetts Institute of Technology
Department of Biological Engineering
77 Massachusetts Avenue
Cambridge, MA 02139, USA
e-mail: *rdkamm@mit.edu*

Ruti Kapon
Weizmann Institute of Science
Department of Biological Chemistry
Rehovot 76100, Israel
e-mail: *ruti.kapon@weizmann.ac.il*

Josef Käs
University of Leipzig
Institute of Experimental Physics I
Division of Soft Matter Physics
Linnéstr. 5
04103 Leipzig, Germany
e-mail: *jkaes@physik.uni-leipzig.de*

Horacio Kido
University of California at Irvine
Mechanical and Aerospace Engineering
Irvine, CA, USA
e-mail: *hkido@uci.edu*

Tobias Kießling
University of Leipzig
Institute of Experimental Physics I
Division of Soft Matter Physics
Linnéstr. 5
04103 Leipzig, Germany
e-mail: *Tobias.Kiessling@uni-leipzig.de*

Jitae Kim
University of California at Irvine
Department of Mechanical
and Aerospace Engineering
Irvine, CA, USA
e-mail: *jitaekim@uci.edu*

Jongbaeg Kim
Yonsei University
School of Mechanical Engineering
1st Engineering Bldg.
Seoul, 120-749, South Korea
e-mail: *kimjb@yonsei.ac.kr*

Nahui Kim
Samsung Advanced Institute of Technology
Research and Development
Seoul, South Korea
e-mail: *nahui.kim@samsung.com*

Kerstin Koch
Rhine-Waal University of Applied Science
Department of Life Science, Biology
and Nanobiotechnology
Landwehr 4
47533 Kleve, Germany
e-mail: *kerstin.koch@hochschule.rhein-waal.de*

Jing Kong
Massachusetts Institute of Technology
Department of Electrical Engineering
and Computer Science
Cambridge, MA, USA
e-mail: *jingkong@mit.edu*

Tobias Kraus
Leibniz-Institut für Neue Materialien gGmbH
Campus D2 2
66123 Saarbrücken, Germany
e-mail: *tobias.kraus@inm-gmbh.de*

Anders Kristensen
Technical University of Denmark
DTU Nanotech
2800 Kongens Lyngby, Denmark
e-mail: *anders.kristensen@nanotech.dtu.dk*

Ratnesh Lal
University of Chicago
Center for Nanomedicine
5841 S Maryland Av
Chicago, IL 60637, USA
e-mail: *rlal@uchicago.edu*

Jan Lammerding
Harvard Medical School
Brigham and Women's Hospital
65 Landsdowne St
Cambridge, MA 02139, USA
e-mail: *jlammerding@rics.bwh.harvard.edu*

Hans Peter Lang
University of Basel
Institute of Physics, National Competence Center
for Research in Nanoscale Science (NCCR) Basel
Klingelbergstr. 82
4056 Basel, Switzerland
e-mail: *hans-peter.lang@unibas.ch*

Carmen LaTorre
Owens Corning Science and Technology
Roofing and Asphalt
2790 Columbus Road
Granville, OH 43023, USA
e-mail: *carmen.latorre@owenscorning.com*

Christophe Laurent
Université Paul Sabatier
CIRIMAT UMR 5085 CNRS
118 Route de Narbonne
31062 Toulouse, France
e-mail: *laurent@chimie.ups-tlse.fr*

Abraham P. Lee
University of California Irvine
Department of Biomedical Engineering
Department of Mechanical
and Aerospace Engineering
Irvine, CA 92697, USA
e-mail: *aplee@uci.edu*

Stephen C. Lee
Ohio State University
Biomedical Engineering Center
Columbus, OH 43210, USA
e-mail: *lee@bme.ohio-state.edu*

Wayne R. Leifert
Adelaide Business Centre
CSIRO Human Nutrition
Adelaide SA 5000, Australia
e-mail: *wayne.leifert@csiro.au*

Liwei Lin
UC Berkeley
Mechanical Engineering Department
5126 Etcheverry
Berkeley, CA 94720-1740, USA
e-mail: *lwlin@me.berkeley.edu*

Yu-Ming Lin
IBM T.J. Watson Research Center
Nanometer Scale Science & Technology
1101 Kitchawan Road
Yorktown Heigths, NY 10598, USA
e-mail: *yming@us.ibm.com*

Marc J. Madou
University of California Irvine
Department of Mechanical and Aerospace
and Biomedical Engineering
Irvine, CA, USA
e-mail: *mmadou@uci.edu*

Othmar Marti
Ulm University
Institute of Experimental Physics
Albert-Einstein-Allee 11
89069 Ulm, Germany
e-mail: *othmar.marti@uni-ulm.de*

Jack Martin
66 Summer Street
Foxborough, MA 02035, USA
e-mail: *jack.martin@alumni.tufts.edu*

Shinji Matsui
University of Hyogo
Laboratory of Advanced Science
and Technology for Industry
Hyogo, Japan
e-mail: *matsui@lasti.u-hyogo.ac.jp*

Mehran Mehregany
Case Western Reserve University
Department of Electrical Engineering
and Computer Science
Cleveland, OH 44106, USA
e-mail: mxm31@cwru.edu

Etienne Menard
Semprius, Inc.
4915 Prospectus Dr.
Durham, NC 27713, USA
e-mail: etienne.menard@semprius.com

Ernst Meyer
University of Basel
Institute of Physics
Basel, Switzerland
e-mail: ernst.meyer@unibas.ch

Robert Modliński
Baolab Microsystems
Terrassa 08220, Spain
e-mail: rmodlinski@gmx.com

Mohammad Mofrad
University of California, Berkeley
Department of Bioengineering
Berkeley, CA 94720, USA
e-mail: mofrad@berkeley.edu

Marc Monthioux
CEMES – UPR A-8011 CNRS
Carbones et Matériaux Carbonés,
Carbons and Carbon-Containing Materials
29 Rue Jeanne Marvig
31055 Toulouse 4, France
e-mail: monthiou@cemes.fr

Markus Morgenstern
RWTH Aachen University
II. Institute of Physics B and JARA-FIT
52056 Aachen, Germany
e-mail: mmorgens@physik.rwth-aachen.de

Seizo Morita
Osaka University
Department of Electronic Engineering
Suita-City
Osaka, Japan
e-mail: smorita@ele.eng.osaka-u.ac.jp

Koichi Mukasa
Hokkaido University
Nanoelectronics Laboratory
Sapporo, Japan
e-mail: mukasa@nano.eng.hokudai.ac.jp

Bradley J. Nelson
Swiss Federal Institute of Technology (ETH)
Institute of Robotics and Intelligent Systems
8092 Zurich, Switzerland
e-mail: bnelson@ethz.ch

Michael Nosonovsky
University of Wisconsin-Milwaukee
Department of Mechanical Engineering
3200 N. Cramer St.
Milwaukee, WI 53211, USA
e-mail: nosonovs@uwm.edu

Hiroshi Onishi
Kanagawa Academy of Science and Technology
Surface Chemistry Laboratory
Kanagawa, Japan
e-mail: oni@net.ksp.or.jp

Alain Peigney
Centre Inter-universitaire de Recherche
sur l'Industrialisation des Matériaux (CIRIMAT)
Toulouse 4, France
e-mail: peigney@chimie.ups-tlse.fr

Oliver Pfeiffer
Individual Computing GmbH
Ingelsteinweg 2d
4143 Dornach, Switzerland
e-mail: oliver.pfeiffer@gmail.com

Haralampos Pozidis
IBM Zurich Research Laboratory
Storage Technologies
Rüschlikon, Switzerland
e-mail: hap@zurich.ibm.com

Robert Puers
Katholieke Universiteit Leuven
ESAT/MICAS
Leuven, Belgium
e-mail: bob.puers@esat.kuleuven.ac.be

Calvin F. Quate
Stanford University
Edward L. Ginzton Laboratory
450 Via Palou
Stanford, CA 94305-4088, USA
e-mail: *quate@stanford.edu*

Oded Rabin
University of Maryland
Department of Materials Science and Engineering
College Park, MD, USA
e-mail: *oded@umd.edu*

Françisco M. Raymo
University of Miami
Department of Chemistry
1301 Memorial Drive
Coral Gables, FL 33146-0431, USA
e-mail: *fraymo@miami.edu*

Manitra Razafinimanana
University of Toulouse III (Paul Sabatier)
Centre de Physique des Plasmas
et leurs Applications (CPPAT)
Toulouse, France
e-mail: *razafinimanana@cpat.ups-tlse.fr*

Ziv Reich
Weizmann Institute of Science Ha'Nesi Ha'Rishon
Department of Biological Chemistry
Rehovot 76100, Israel
e-mail: *ziv.reich@weizmann.ac.il*

John A. Rogers
University of Illinois
Department of Materials Science and Engineering
Urbana, IL, USA
e-mail: *jrogers@uiuc.edu*

Cosmin Roman
ETH Zurich
Micro and Nanosystems Department of Mechanical
and Process Engineering
8092 Zurich, Switzerland
e-mail: *cosmin.roman@micro.mavt.ethz.ch*

Marina Ruths
University of Massachusetts Lowell
Department of Chemistry
1 University Avenue
Lowell, MA 01854, USA
e-mail: *marina_ruths@uml.edu*

Ozgur Sahin
The Rowland Institute at Harvard
100 Edwin H. Land Blvd
Cambridge, MA 02142, USA
e-mail: *sahin@rowland.harvard.edu*

Akira Sasahara
Japan Advanced Institute
of Science and Technology
School of Materials Science
1-1 Asahidai
923-1292 Nomi, Japan
e-mail: *sasahara@jaist.ac.jp*

Helmut Schift
Paul Scherrer Institute
Laboratory for Micro- and Nanotechnology
5232 Villigen PSI, Switzerland
e-mail: *helmut.schift@psi.ch*

André Schirmeisen
University of Münster
Institute of Physics
Wilhelm-Klemm-Str. 10
48149 Münster, Germany
e-mail: *schirmeisen@uni-muenster.de*

Christian Schulze
Beiersdorf AG
Research & Development
Unnastr. 48
20245 Hamburg, Germany
e-mail: *christian.schulze@beiersdorf.com;
christian.schulze@uni-leipzig.de*

Alexander Schwarz
University of Hamburg
Institute of Applied Physics
Jungiusstr. 11
20355 Hamburg, Germany
e-mail: *aschwarz@physnet.uni-hamburg.de*

Udo D. Schwarz
Yale University
Department of Mechanical Engineering
15 Prospect Street
New Haven, CT 06520-8284, USA
e-mail: *udo.schwarz@yale.edu*

Philippe Serp
Ecole Nationale Supérieure d'Ingénieurs
en Arts Chimiques et Technologiques
Laboratoire de Chimie de Coordination (LCC)
118 Route de Narbonne
31077 Toulouse, France
e-mail: *philippe.serp@ensiacet.fr*

Huamei (Mary) Shang
GE Healthcare
4855 W. Electric Ave.
Milwaukee, WI 53219, USA
e-mail: *huamei.shang@ge.com*

Susan B. Sinnott
University of Florida
Department of Materials Science and Engineering
154 Rhines Hall
Gainesville, FL 32611-6400, USA
e-mail: *ssinn@mse.ufl.edu*

Anisoara Socoliuc
SPECS Zurich GmbH
Technoparkstr. 1
8005 Zurich, Switzerland
e-mail: *socoliuc@nanonis.com*

Olav Solgaard
Stanford University
E.L. Ginzton Laboratory
450 Via Palou
Stanford, CA 94305-4088, USA
e-mail: *solgaard@stanford.edu*

Dan Strehle
University of Leipzig
Institute of Experimental Physics I
Division of Soft Matter Physics
Linnéstr. 5
04103 Leipzig, Germany
e-mail: *dan.strehle@uni-leipzig.de*

Carsten Stüber
University of Leipzig
Institute of Experimental Physics I
Division of Soft Matter Physics
Linnéstr. 5
04103 Leipzig, Germany
e-mail: *stueber@rz.uni-leipzig.de*

Yu-Chuan Su
ESS 210
Department of Engineering and System Science 101
Kuang-Fu Road
Hsinchu, 30013, Taiwan
e-mail: *ycsu@ess.nthu.edu.tw*

Kazuhisa Sueoka
Graduate School of Information Science
and Technology
Hokkaido University
Nanoelectronics Laboratory
Kita-14, Nishi-9, Kita-ku
060-0814 Sapporo, Japan
e-mail: *sueoka@nano.isthokudai.ac.jp*

Yasuhiro Sugawara
Osaka University
Department of Applied Physics
Yamada-Oka 2-1, Suita
565-0871 Osaka, Japan
e-mail: *sugawara@ap.eng.osaka-u.ac.jp*

Benjamin Sullivan
TearLab Corp.
11025 Roselle Street
San Diego, CA 92121, USA
e-mail: *bdsulliv@TearLab.com*

Paul Swanson
Nexogen, Inc.
Engineering
8360 C Camino Santa Fe
San Diego, CA 92121, USA
e-mail: *pswanson@nexogentech.com*

Yung-Chieh Tan
Washington University School of Medicine
Department of Medicine
Division of Dermatology
660 S. Euclid Ave.
St. Louis, MO 63110, USA
e-mail: *ytanster@gmail.com*

Shia-Yen Teh
University of California at Irvine
Biomedical Engineering Department
3120 Natural Sciences II
Irvine, CA 92697-2715, USA
e-mail: *steh@uci.edu*

W. Merlijn van Spengen
Leiden University
Kamerlingh Onnes Laboratory
Niels Bohrweg 2
Leiden, CA 2333, The Netherlands
e-mail: *spengen@physics.leidenuniv.nl*

Peter Vettiger
University of Neuchâtel
SAMLAB
Jaquet-Droz 1
2002 Neuchâtel, Switzerland
e-mail: *peter.vettiger@unine.ch*

Franziska Wetzel
University of Leipzig
Institute of Experimental Physics I
Division of Soft Matter Physics
Linnéstr. 5
04103 Leipzig, Germany
e-mail: *franziska.wetzel@uni-leipzig.de*

Heiko Wolf
IBM Research GmbH
Zurich Research Laboratory
Säumerstr. 4
8803 Rüschlikon, Switzerland
e-mail: *hwo@zurich.ibm.com*

Darrin J. Young
Case Western Reserve University
Department of EECS, Glennan 510
10900 Euclid Avenue
Cleveland, OH 44106, USA
e-mail: *djy@po.cwru.edu*

Babak Ziaie
Purdue University
Birck Nanotechnology Center
1205 W. State St.
West Lafayette, IN 47907-2035, USA
e-mail: *bziaie@purdue.edu*

Christian A. Zorman
Case Western Reserve University
Department of Electrical Engineering
and Computer Science
10900 Euclid Avenue
Cleveland, OH 44106, USA
e-mail: *caz@case.edu*

Jim V. Zoval
Saddleback College
Department of Math and Science
28000 Marguerite Parkway
Mission Viejo, CA 92692, USA
e-mail: *jzoval@saddleback.edu*

目 录

缩略语

Part E 厚膜分子润滑

38. 超薄非晶碳薄膜的纳米摩擦技术 ... 3
38.1 常见沉积技术描述 ... 7
38.2 化学和物理镀膜的特征 ... 11
38.3 微机械和摩擦学镀膜的特征 ... 17
38.4 结 论 ... 38
参考文献 ... 39

39. 纳米摩擦技术和表面保护的自组装单分子膜 ... 43
39.1 背 景 ... 43
39.2 有机化学入门 ... 47
39.3 自组装单分子膜：分子链中的基片、隔链、端基 ... 50
39.4 自组装单分子膜的接触角和纳米摩擦特性 ... 53
39.5 总 结 ... 74
参考文献 ... 76

40. 纳米边界润滑研究 ... 81
40.1 边界薄膜 ... 81
40.2 纳米形变、分子结构、外延和纳米摩擦研究 ... 82
40.3 新兴全氟聚醚润滑薄膜的纳米摩擦学、电学、化学降解研究和环境影响 ... 100
40.4 离子液体薄膜的纳米摩擦学和电学研究 ... 109
40.5 结 论 ... 126
参考文献 ... 127

Contents

List of Abbreviations

Part E Molecularly Thick Films for Lubrication

38 Nanotribology of Ultrathin and Hard Amorphous Carbon Films
Bharat Bhushan 1269
- 38.1 Description of Common Deposition Techniques 1273
- 38.2 Chemical and Physical Coating Characterization 1277
- 38.3 Micromechanical and Tribological Coating Characterization 1283
- 38.4 Closure 1304
- **References** 1305

39 Self-Assembled Monolayers for Nanotribology and Surface Protection
Bharat Bhushan 1309
- 39.1 Background 1309
- 39.2 A Primer to Organic Chemistry 1313
- 39.3 Self-Assembled Monolayers: Substrates, Spacer Chains, and End Groups in the Molecular Chains 1316
- 39.4 Contact Angle and Nanotribological Properties of SAMs 1319
- 39.5 Summary 1340
- **References** 1342

40 Nanoscale Boundary Lubrication Studies
Bharat Bhushan 1347
- 40.1 Boundary Films 1347
- 40.2 Nanodeformation, Molecular Conformation, Spreading, and Nanotribological Studies 1348
- 40.3 Nanotribological, Electrical, and Chemical Degradations Studies and Environmental Effects in Novel PFPE Lubricant Films 1366
- 40.4 Nanotribological and Electrical Studies of Ionic Liquid Films 1375
- 40.5 Conclusions 1392
- **References** 1393

List of Abbreviations

μCP	microcontact printing
1-D	one-dimensional
18-MEA	18-methyl eicosanoic acid
2-D	two-dimensional
2-DEG	two-dimensional electron gas
3-APTES	3-aminopropyltriethoxysilane
3-D	three-dimensional

A

a-BSA	anti-bovine serum albumin
a-C	amorphous carbon
A/D	analog-to-digital
AA	amino acid
AAM	anodized alumina membrane
ABP	actin binding protein
AC	alternating-current
AC	amorphous carbon
ACF	autocorrelation function
ADC	analog-to-digital converter
ADXL	analog devices accelerometer
AFAM	atomic force acoustic microscopy
AFM	atomic force microscope
AFM	atomic force microscopy
AKD	alkylketene dimer
ALD	atomic layer deposition
AM	amplitude modulation
AMU	atomic mass unit
AOD	acoustooptical deflector
AOM	acoustooptical modulator
AP	alkaline phosphatase
APB	actin binding protein
APCVD	atmospheric-pressure chemical vapor deposition
APDMES	aminopropyldimethylethoxysilane
APTES	aminopropyltriethoxysilane
ASIC	application-specific integrated circuit
ASR	analyte-specific reagent
ATP	adenosine triphosphate

B

BAP	barometric absolute pressure
BAPDMA	behenyl amidopropyl dimethylamine glutamate
bcc	body-centered cubic
BCH	brucite-type cobalt hydroxide
BCS	Bardeen–Cooper–Schrieffer
BD	blu-ray disc
BDCS	biphenyldimethylchlorosilane
BE	boundary element
BFP	biomembrane force probe
BGA	ball grid array
BHF	buffered HF
BHPET	1,1'-(3,6,9,12,15-pentaoxapentadecane-1,15-diyl)bis(3-hydroxyethyl-1H-imidazolium-1-yl) di[bis(trifluoromethanesulfonyl)imide]
BHPT	1,1'-(pentane-1,5-diyl)bis(3-hydroxyethyl-1H-imidazolium-1-yl) di[bis(trifluoromethanesulfonyl)imide]
BiCMOS	bipolar CMOS
bioMEMS	biomedical microelectromechanical system
bioNEMS	biomedical nanoelectromechanical system
BMIM	1-butyl-3-methylimidazolium
BP	bit pitch
BPAG1	bullous pemphigoid antigen 1
BPT	biphenyl-4-thiol
BPTC	cross-linked BPT
BSA	bovine serum albumin
BST	barium strontium titanate
BTMAC	behentrimonium chloride

C

CA	constant amplitude
CA	contact angle
CAD	computer-aided design
CAH	contact angle hysteresis
cAMP	cyclic adenosine monophosphate
CAS	Crk-associated substrate
CBA	cantilever beam array
CBD	chemical bath deposition
CCD	charge-coupled device
CCVD	catalytic chemical vapor deposition
CD	compact disc
CD	critical dimension
CDR	complementarity determining region
CDW	charge density wave
CE	capillary electrophoresis
CE	constant excitation
CEW	continuous electrowetting
CG	controlled geometry
CHO	Chinese hamster ovary
CIC	cantilever in cantilever
CMC	cell membrane complex
CMC	critical micelle concentration
CMOS	complementary metal–oxide–semiconductor
CMP	chemical mechanical polishing

CNF	carbon nanofiber	DOS	density of states
CNFET	carbon nanotube field-effect transistor	DP	decylphosphonate
CNT	carbon nanotube	DPN	dip-pen nanolithography
COC	cyclic olefin copolymer	DRAM	dynamic random-access memory
COF	chip-on-flex	DRIE	deep reactive ion etching
COF	coefficient of friction	ds	double-stranded
COG	cost of goods	DSC	differential scanning calorimetry
CoO	cost of ownership	DSP	digital signal processor
COS	CV-1 in origin with SV40	DTR	discrete track recording
CP	circularly permuted	DTSSP	3,3'-dithio-bis(sulfosuccinimidylproprionate)
CPU	central processing unit		
CRP	C-reactive protein	DUV	deep-ultraviolet
CSK	cytoskeleton	DVD	digital versatile disc
CSM	continuous stiffness measurement	DWNT	double-walled CNT
CTE	coefficient of thermal expansion		
Cu-TBBP	Cu-tetra-3,5 di-tertiary-butyl-phenyl porphyrin		

E

EAM	embedded atom method
EB	electron beam
EBD	electron beam deposition
EBID	electron-beam-induced deposition
EBL	electron-beam lithography
ECM	extracellular matrix
ECR-CVD	electron cyclotron resonance chemical vapor deposition
ED	electron diffraction
EDC	1-ethyl-3-(3-diamethylaminopropyl) carbodiimide
EDL	electrostatic double layer
EDP	ethylene diamine pyrochatechol
EDTA	ethylenediamine tetraacetic acid
EDX	energy-dispersive x-ray
EELS	electron energy loss spectra
EFM	electric field gradient microscopy
EFM	electrostatic force microscopy
EHD	elastohydrodynamic
EO	electroosmosis
EOF	electroosmotic flow
EOS	electrical overstress
EPA	Environmental Protection Agency
EPB	electrical parking brake
ESD	electrostatic discharge
ESEM	environmental scanning electron microscope
EU	European Union
EUV	extreme ultraviolet
EW	electrowetting
EWOD	electrowetting on dielectric

D

DBR	distributed Bragg reflector
DC-PECVD	direct-current plasma-enhanced CVD
DC	direct-current
DDT	dichlorodiphenyltrichloroethane
DEP	dielectrophoresis
DFB	distributed feedback
DFM	dynamic force microscopy
DFS	dynamic force spectroscopy
DGU	density gradient ultracentrifugation
DI	FESPdigital instrument force modulation etched Si probe
DI	TESPdigital instrument tapping mode etched Si probe
DI	digital instrument
DI	deionized
DIMP	diisopropylmethylphosphonate
DIP	dual inline packaging
DIPS	industrial postpackaging
DLC	diamondlike carbon
DLP	digital light processing
DLVO	Derjaguin–Landau–Verwey–Overbeek
DMD	deformable mirror display
DMD	digital mirror device
DMDM	1,3-dimethylol-5,5-dimethyl
DMMP	dimethylmethylphosphonate
DMSO	dimethyl sulfoxide
DMT	Derjaguin–Muller–Toporov
DNA	deoxyribonucleic acid
DNT	2,4-dinitrotoluene
DOD	Department of Defense
DOE	Department of Energy
DOE	diffractive optical element
DOF	degree of freedom
DOPC	1,2-dioleoyl-sn-glycero-3-phosphocholine

F

F-actin	filamentous actin
FA	focal adhesion
FAA	formaldehyde–acetic acid–ethanol
FACS	fluorescence-activated cell sorting

FAK	focal adhesion kinase		HDT	hexadecanethiol
FBS	fetal bovine serum		HDTV	high-definition television
FC	flip-chip		HEK	human embryonic kidney 293
FCA	filtered cathodic arc		HEL	hot embossing lithography
fcc	face-centered cubic		HEXSIL	hexagonal honeycomb polysilicon
FCP	force calibration plot		HF	hydrofluoric
FCS	fluorescence correlation spectroscopy		HMDS	hexamethyldisilazane
FD	finite difference		HNA	hydrofluoric-nitric-acetic
FDA	Food and Drug Administration		HOMO	highest occupied molecular orbital
FE	finite element		HOP	highly oriented pyrolytic
FEM	finite element method		HOPG	highly oriented pyrolytic graphite
FEM	finite element modeling		HOT	holographic optical tweezer
FESEM	field emission SEM		HP	hot-pressing
FESP	force modulation etched Si probe		HPI	hexagonally packed intermediate
FET	field-effect transistor		HRTEM	high-resolution transmission electron microscope
FFM	friction force microscope		HSA	human serum albumin
FFM	friction force microscopy		HtBDC	hexa-*tert*-butyl-decacyclene
FIB-CVD	focused ion beam chemical vapor deposition		HTCS	high-temperature superconductivity
FIB	focused ion beam		HTS	high throughput screening
FIM	field ion microscope		HUVEC	human umbilical venous endothelial cell
FIP	feline coronavirus			
FKT	Frenkel–Kontorova–Tomlinson		**I**	
FM	frequency modulation			
FMEA	failure-mode effect analysis		IBD	ion beam deposition
FP6	Sixth Framework Program		IC	integrated circuit
FP	fluorescence polarization		ICA	independent component analysis
FPR	*N*-formyl peptide receptor		ICAM-1	intercellular adhesion molecules 1
FS	force spectroscopy		ICAM-2	intercellular adhesion molecules 2
FTIR	Fourier-transform infrared		ICT	information and communication technology
FV	force–volume		IDA	interdigitated array
			IF	intermediate filament
G			IF	intermediate-frequency
			IFN	interferon
GABA	γ-aminobutyric acid		IgG	immunoglobulin G
GDP	guanosine diphosphate		IKVAV	isoleucine–lysine–valine–alanine–valine
GF	gauge factor		IL	ionic liquid
GFP	green fluorescent protein		IMAC	immobilized metal ion affinity chromatography
GMR	giant magnetoresistive		IMEC	Interuniversity MicroElectronics Center
GOD	glucose oxidase		IR	infrared
GPCR	G-protein coupled receptor		ISE	indentation size effect
GPS	global positioning system		ITO	indium tin oxide
GSED	gaseous secondary-electron detector		ITRS	International Technology Roadmap for Semiconductors
GTP	guanosine triphosphate		IWGN	Interagency Working Group on Nanoscience, Engineering, and Technology
GW	Greenwood and Williamson			
			J	
H				
			JC	jump-to-contact
HAR	high aspect ratio		JFIL	jet-and-flash imprint lithography
HARMEMS	high-aspect-ratio MEMS		JKR	Johnson–Kendall–Roberts
HARPSS	high-aspect-ratio combined poly- and single-crystal silicon			
HBM	human body model			
hcp	hexagonal close-packed			
HDD	hard-disk drive			

K

KASH	Klarsicht, ANC-1, Syne Homology
KPFM	Kelvin probe force microscopy

L

LA	lauric acid
LAR	low aspect ratio
LB	Langmuir–Blodgett
LBL	layer-by-layer
LCC	leadless chip carrier
LCD	liquid-crystal display
LCoS	liquid crystal on silicon
LCP	liquid-crystal polymer
LDL	low-density lipoprotein
LDOS	local density of states
LED	light-emitting diode
LFA-1	leukocyte function-associated antigen-1
LFM	lateral force microscope
LFM	lateral force microscopy
LIGA	Lithographie Galvanoformung Abformung
LJ	Lennard-Jones
LMD	laser microdissection
LMPC	laser microdissection and pressure catapulting
LN	liquid-nitrogen
LoD	limit-of-detection
LOR	lift-off resist
LPC	laser pressure catapulting
LPCVD	low-pressure chemical vapor deposition
LSC	laser scanning cytometry
LSN	low-stress silicon nitride
LT-SFM	low-temperature scanning force microscope
LT-SPM	low-temperature scanning probe microscopy
LT-STM	low-temperature scanning tunneling microscope
LT	low-temperature
LTM	laser tracking microrheology
LTO	low-temperature oxide
LTRS	laser tweezers Raman spectroscopy
LUMO	lowest unoccupied molecular orbital
LVDT	linear variable differential transformer

M

MALDI	matrix assisted laser desorption ionization
MAP	manifold absolute pressure
MAPK	mitogen-activated protein kinase
MAPL	molecular assembly patterning by lift-off
MBE	molecular-beam epitaxy
MC	microcantilever
MC	microcapillary
MCM	multi-chip module
MD	molecular dynamics
ME	metal-evaporated
MEMS	microelectromechanical system
MExFM	magnetic exchange force microscopy
MFM	magnetic field microscopy
MFM	magnetic force microscope
MFM	magnetic force microscopy
MHD	magnetohydrodynamic
MIM	metal–insulator–metal
MIMIC	micromolding in capillaries
MLE	maximum likelihood estimator
MOCVD	metalorganic chemical vapor deposition
MOEMS	microoptoelectromechanical system
MOS	metal–oxide–semiconductor
MOSFET	metal–oxide–semiconductor field-effect transistor
MP	metal particle
MPTMS	mercaptopropyltrimethoxysilane
MRFM	magnetic resonance force microscopy
MRFM	molecular recognition force microscopy
MRI	magnetic resonance imaging
MRP	molecular recognition phase
MscL	mechanosensitive channel of large conductance
MST	microsystem technology
MT	microtubule
mTAS	micro total analysis system
MTTF	mean time to failure
MUMP	multiuser MEMS process
MVD	molecular vapor deposition
MWCNT	multiwall carbon nanotube
MWNT	multiwall nanotube
MYD/BHW	Muller–Yushchenko–Derjaguin/Burgess–Hughes–White

N

NA	numerical aperture
NADIS	nanoscale dispensing
NASA	National Aeronautics and Space Administration
NC-AFM	noncontact atomic force microscopy
NEMS	nanoelectromechanical system
NGL	next-generation lithography
NHS	N-hydroxysuccinimidyl
NIH	National Institute of Health
NIL	nanoimprint lithography
NIST	National Institute of Standards and Technology
NMP	no-moving-part
NMR	nuclear magnetic resonance
NMR	nuclear mass resonance
NNI	National Nanotechnology Initiative

NOEMS	nanooptoelectromechanical system		PET	poly(ethyleneterephthalate)
NP	nanoparticle		PETN	pentaerythritol tetranitrate
NP	nanoprobe		PFDA	perfluorodecanoic acid
NSF	National Science Foundation		PFDP	perfluorodecylphosphonate
NSOM	near-field scanning optical microscopy		PFDTES	perfluorodecyltriethoxysilane
NSTC	National Science and Technology Council		PFM	photonic force microscope
			PFOS	perfluorooctanesulfonate
NTA	nitrilotriacetate		PFPE	perfluoropolyether
nTP	nanotransfer printing		PFTS	perfluorodecyltricholorosilane
			PhC	photonic crystal
			PI3K	phosphatidylinositol-3-kinase

O

			PI	polyisoprene
			PID	proportional–integral–differential
ODA	octadecylamine		PKA	protein kinase
ODDMS	n-octadecyldimethyl(dimethylamino)silane		PKC	protein kinase C
			PKI	protein kinase inhibitor
ODMS	n-octyldimethyl(dimethylamino)silane		PL	photolithography
ODP	octadecylphosphonate		PLC	phospholipase C
ODTS	octadecyltrichlorosilane		PLD	pulsed laser deposition
OLED	organic light-emitting device		PMAA	poly(methacrylic acid)
OM	optical microscope		PML	promyelocytic leukemia
OMVPE	organometallic vapor-phase epitaxy		PMMA	poly(methyl methacrylate)
OS	optical stretcher		POCT	point-of-care testing
OT	optical tweezers		POM	polyoxy-methylene
OTRS	optical tweezers Raman spectroscopy		PP	polypropylene
OTS	octadecyltrichlorosilane		PPD	p-phenylenediamine
oxLDL	oxidized low-density lipoprotein		PPMA	poly(propyl methacrylate)
			PPy	polypyrrole
			PS-PDMS	poly(styrene-b-dimethylsiloxane)
			PS/clay	polystyrene/nanoclay composite

P

			PS	polystyrene
P–V	peak-to-valley		PSA	prostate-specific antigen
PAA	poly(acrylic acid)		PSD	position-sensitive detector
PAA	porous anodic alumina		PSD	position-sensitive diode
PAH	poly(allylamine hydrochloride)		PSD	power-spectral density
PAPP	p-aminophenyl phosphate		PSG	phosphosilicate glass
Pax	paxillin		PSGL-1	P-selectin glycoprotein ligand-1
PBC	periodic boundary condition		PTFE	polytetrafluoroethylene
PBS	phosphate-buffered saline		PUA	polyurethane acrylate
PC	polycarbonate		PUR	polyurethane
PCB	printed circuit board		PVA	polyvinyl alcohol
PCL	polycaprolactone		PVD	physical vapor deposition
PCR	polymerase chain reaction		PVDC	polyvinylidene chloride
PDA	personal digital assistant		PVDF	polyvinyledene fluoride
PDMS	polydimethylsiloxane		PVS	polyvinylsiloxane
PDP	2-pyridyldithiopropionyl		PWR	plasmon-waveguide resonance
PDP	pyridyldithiopropionate		PZT	lead zirconate titanate
PE	polyethylene			
PECVD	plasma-enhanced chemical vapor deposition			

Q

PEEK	polyetheretherketone		QB	quantum box
PEG	polyethylene glycol		QCM	quartz crystal microbalance
PEI	polyethyleneimine		QFN	quad flat no-lead
PEN	polyethylene naphthalate		QPD	quadrant photodiode
PES	photoemission spectroscopy		QWR	quantum wire
PES	position error signal			

R

RBC	red blood cell
RCA	Radio Corporation of America
RF	radiofrequency
RFID	radiofrequency identification
RGD	arginine–glycine–aspartic
RH	relative humidity
RHEED	reflection high-energy electron diffraction
RICM	reflection interference contrast microscopy
RIE	reactive-ion etching
RKKY	Ruderman–Kittel–Kasuya–Yoshida
RMS	root mean square
RNA	ribonucleic acid
ROS	reactive oxygen species
RPC	reverse phase column
RPM	revolutions per minute
RSA	random sequential adsorption
RT	room temperature
RTP	rapid thermal processing

S

SAE	specific adhesion energy
SAM	scanning acoustic microscopy
SAM	self-assembled monolayer
SARS-CoV	syndrome associated coronavirus
SATI	self-assembly, transfer, and integration
SATP	(S-acetylthio)propionate
SAW	surface acoustic wave
SB	Schottky barrier
SCFv	single-chain fragment variable
SCM	scanning capacitance microscopy
SCPM	scanning chemical potential microscopy
SCREAM	single-crystal reactive etching and metallization
SDA	scratch drive actuator
SEcM	scanning electrochemical microscopy
SEFM	scanning electrostatic force microscopy
SEM	scanning electron microscope
SEM	scanning electron microscopy
SFA	surface forces apparatus
SFAM	scanning force acoustic microscopy
SFD	shear flow detachment
SFIL	step and flash imprint lithography
SFM	scanning force microscope
SFM	scanning force microscopy
SGS	small-gap semiconducting
SICM	scanning ion conductance microscopy
SIM	scanning ion microscope
SIP	single inline package
SKPM	scanning Kelvin probe microscopy
SL	soft lithography
SLIGA	sacrificial LIGA
SLL	sacrificial layer lithography
SLM	spatial light modulator
SMA	shape memory alloy
SMM	scanning magnetic microscopy
SNOM	scanning near field optical microscopy
SNP	single nucleotide polymorphisms
SNR	signal-to-noise ratio
SOG	spin-on-glass
SOI	silicon-on-insulator
SOIC	small outline integrated circuit
SoS	silicon-on-sapphire
SP-STM	spin-polarized STM
SPM	scanning probe microscope
SPM	scanning probe microscopy
SPR	surface plasmon resonance
sPROM	structurally programmable microfluidic system
SPS	spark plasma sintering
SRAM	static random access memory
SRC	sampling rate converter
SSIL	step-and-stamp imprint lithography
SSRM	scanning spreading resistance microscopy
STED	stimulated emission depletion
SThM	scanning thermal microscope
STM	scanning tunneling microscope
STM	scanning tunneling microscopy
STORM	statistical optical reconstruction microscopy
STP	standard temperature and pressure
STS	scanning tunneling spectroscopy
SUN	Sad1p/UNC-84
SWCNT	single-wall carbon nanotube
SWCNT	single-walled carbon nanotube
SWNT	single wall nanotube
SWNT	single-wall nanotube

T

TA	tilt angle
TASA	template-assisted self-assembly
TCM	tetracysteine motif
TCNQ	tetracyanoquinodimethane
TCP	tricresyl phosphate
TEM	transmission electron microscope
TEM	transmission electron microscopy
TESP	tapping mode etched silicon probe
TGA	thermogravimetric analysis
TI	Texas Instruments
TIRF	total internal reflection fluorescence
TIRM	total internal reflection microscopy
TLP	transmission-line pulse
TM	tapping mode
TMAH	tetramethyl ammonium hydroxide
TMR	tetramethylrhodamine
TMS	tetramethylsilane

TMS	trimethylsilyl
TNT	trinitrotoluene
TP	track pitch
TPE-FCCS	two-photon excitation fluorescence cross-correlation spectroscopy
TPI	threads per inch
TPMS	tire pressure monitoring system
TR	torsional resonance
TREC	topography and recognition
TRIM	transport of ions in matter
TSDC	thermally stimulated depolarization current
TTF	tetrathiafulvalene
TV	television

U

UAA	unnatural AA
UHV	ultrahigh vacuum
ULSI	ultralarge-scale integration
UML	unified modeling language
UNCD	ultrananocrystalline diamond
UV	ultraviolet
UVA	ultraviolet A

V

VBS	vinculin binding site
VCO	voltage-controlled oscillator
VCSEL	vertical-cavity surface-emitting laser
vdW	van der Waals
VHH	variable heavy–heavy
VLSI	very large-scale integration
VOC	volatile organic compound
VPE	vapor-phase epitaxy
VSC	vehicle stability control

X

XPS	x-ray photon spectroscopy
XRD	x-ray powder diffraction

Y

YFP	yellow fluorescent protein

Z

Z-DOL	perfluoropolyether

Part E Molecularly Thick Films for Lubrication

38 Nanotribology of Ultrathin and Hard Amorphous Carbon Films
Bharat Bhushan, Columbus, USA

39 Self-Assembled Monolayers for Nanotribology and Surface Protection
Bharat Bhushan, Columbus, USA

40 Nanoscale Boundary Lubrication Studies
Bharat Bhushan, Columbus, USA

38. Nanotribology of Ultrathin and Hard Amorphous Carbon Films

Bharat Bhushan

One of the best materials to use in applications that require very low wear and reduced friction is diamond, especially in the form of a diamond coating. Unfortunately, true diamond coatings can only be deposited at high temperatures and on selected substrates, and they require surface finishing. However, hard amorphous carbon – commonly known as diamond-like carbon or a DLC coating – has similar mechanical, thermal and optical properties to those of diamond. It can also be deposited at a wide range of thicknesses using a variety of deposition processes on various substrates at or near room temperature. The coatings reproduce the topography of the substrate, removing the need for finishing. The friction and wear properties of some DLC coatings make them very attractive for some tribological applications. The most significant current industrial application of DLC coatings is in magnetic storage devices.

In this chapter, the state-of-the-art in the chemical, mechanical and tribological characterization of ultrathin amorphous carbon coatings is presented.

EELS and Raman spectroscopies can be used to characterize amorphous carbon coatings chemically. The prevailing atomic arrangement in the DLC coatings is amorphous or quasi-amorphous, with small diamond (sp^3), graphite (sp^2) and other unidentifiable micro- or nanocrystallites. Most DLC coatings, except for those produced using a filtered cathodic arc, contain from a few to about 50 at. % hydrogen. Sometimes hydrogen is deliberately incorporated into the sputtered and ion-plated coatings in order to tailor their properties.

Amorphous carbon coatings deposited by different techniques exhibit different mechanical and tribological properties. Thin coatings deposited by filtered cathodic arc, ion beam and ECR-CVD hold

38.1	**Description of Common Deposition Techniques** 1273
	38.1.1 Filtered Cathodic Arc Deposition 1276
	38.1.2 Ion Beam Deposition 1276
	38.1.3 Electron Cyclotron Resonance Chemical Vapor Deposition 1277
	38.1.4 Sputtering Deposition 1277
	38.1.5 Plasma-Enhanced Chemical Vapor Deposition 1277
38.2	**Chemical and Physical Coating Characterization** 1277
	38.2.1 EELS and Raman Spectroscopy 1278
	38.2.2 Hydrogen Concentrations 1280
	38.2.3 Physical Properties 1282
	38.2.4 Summary 1283
38.3	**Micromechanical and Tribological Coating Characterization** 1283
	38.3.1 Micromechanical Characterization . 1283
	38.3.2 Microscratch and Microwear Studies 1293
	38.3.3 Macroscale Tribological Characterization 1298
	38.3.4 Coating Continuity Analysis........... 1303
38.4	**Closure** 1304
References 1305

hold much promise for tribological applications. Coatings of 5 nm or even less provide wear protection. A nanoindenter can be used to measure DLC coating hardness, elastic modulus, fracture toughness and fatigue life. Microscratch and microwear tests can be performed on the coatings using either a nanoindenter or an AFM, and along with accelerated wear testing, can be used to screen potential industrial coatings. For the examples shown in this chapter, the trends observed in such tests were similar to those found in functional tests.

Carbon exists in both crystalline and amorphous forms and exhibits both metallic and nonmetallic characteristics [38.1–3]. Forms of crystalline carbon include graphite, diamond, and a family of fullerenes (Fig. 38.1). The graphite and diamond are infinite periodic network solids with a planar structure, whereas the fullerenes are a molecular form of pure carbon with a finite network and a nonplanar structure. Graphite has a hexagonal, layered structure with weak interlayer bonding forces and it exhibits excellent lubrication properties. The graphite crystal may be visualized as infinite parallel layers of hexagons stacked 0.34 nm apart with an interatomic distance of 0.1415 nm between the carbon atoms in the basal plane. Each atom lying in the basal planes is trigonally coordinated and closely packed with strong σ (covalent) bonds to its three carbon neighbors via hybrid sp^2 orbitals. The fourth electron lies in a p_z orbital lying normal to the σ bonding plane and forms a weak π bond by overlapping side-to-side with a p_z orbital of an adjacent atom to which the carbon is attached by a σ bond. The layers (basal planes) themselves are relatively far apart and the forces that bond them are weak van der Waals forces. These layers can align themselves parallel to the direction of the relative motion and slide over one another with relative ease, meaning low friction. Strong interatomic bonding and packing in each layer is thought to help reduce wear. The operating environment has a significant influence on the lubrication – low friction and low wear – properties of graphite. It lubricates better in a humid environment than a dry one, due to the adsorption of water vapor and other gases from the environment, which further weakens the interlayer bonding forces and results in easy shear and transfer of the crystallite platelets to the mating surface. Thus, transfer plays an important role in controlling friction and wear. Graphite oxidizes at high operating temperatures and can be used up to about 430 °C.

One of the most well-known fullerene molecules is C_{60}, commonly known as buckyballs. Since these C_{60} molecules are very stable and do not require additional atoms to satisfy chemical bonding requirements, they are expected to have low adhesion to the mating surface and low surface energy. Since the C_{60} molecule, which has a perfect spherical symmetry, bonds only weakly to other molecules, C_{60} clusters readily become detached, similar to other layered lattice structures, and either get transferred to the mating surface by mechanical compaction or are present as loose wear particles that may roll like tiny ball bearings in a sliding contact, resulting in low friction and wear. The wear particles are expected to be harder than as-deposited C_{60} molecules, because of their phase transformation at the high-asperity contact pressures present in a sliding interface. The low surface energy, the spherical shapes of C_{60} molecules, the weak intermolecular bonding, and the high load bearing capacity offer vast potential for various mechanical and tribological applications. Sublimed C_{60} coatings and fullerene particles used as an additive to mineral oils and greases have been reported to be good solid lubricants comparable to graphite and MoS_2 [38.4–6].

Diamond crystallizes in a modified face-centered cubic (fcc) structure with an interatomic distance of 0.154 nm. The diamond cubic lattice consists of two interpenetrating fcc lattices displaced by a quarter of the cube diagonal. Each carbon atom is tetrahedrally coordinated, making strong σ (covalent) bonds to its four carbon neighbors using the hybrid sp^3 atomic orbitals, which accounts for it having the highest hardness (80–104 GPa) and thermal conductivity (900–2100 W/(m K)), on the order of five times that of

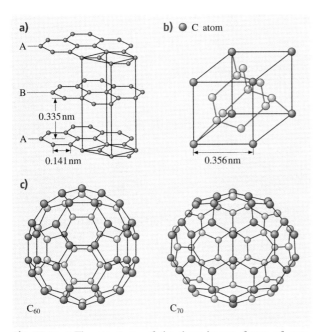

Fig. 38.1a–c The structures of the three known forms of crystalline carbon: (**a**) hexagonal structure of graphite, (**b**) modified face-centered cubic (fcc) structure (two interpenetrating fcc lattices displaced by a quarter of the cube diagonal) of diamond (each atom is bonded to four others that form the corners of a tetrahedron), and (**c**) the structures of the two most common fullerenes: a soccer ball C_{60} and a rugby ball C_{70} molecules

copper) of any known solid, as well as high electrical resistivity and optical transmission and a large optical band gap. It is relatively chemically inert, and it exhibits poor adhesion with other solids, enhancing its low friction and wear properties. Its high thermal conductivity permits the dissipation of frictional heat during sliding and it protects the interface, and the dangling carbon bonds on the surface react with the environment to form hydrocarbons that act as good lubrication films. These are some of the reasons for the low friction and wear of the diamond. Diamond and its coatings find many industrial applications: tribological applications (low friction and wear), optical applications (exceptional optical transmission, high abrasion resistance), and thermal management or heat sink applications (high thermal conductivity). Diamond can be used at high temperatures; it starts to graphitize at about 1000 °C in ambient air and at about 1400 °C in vacuum. Diamond is an attractive material for cutting tools, abrasives for grinding wheels and lapping compounds, and other extreme wear applications.

Natural diamond – particularly in large quantities – is very expensive, and so diamond coatings – a low-cost alternative – are attractive. True diamond coatings are deposited by chemical vapor deposition (CVD) processes at high substrate temperatures (on the order of 800 °C). They adhere best on silicon substrate and require an interlayer for other substrates. One major hindrance to the widespread use of true diamond films in tribological, optical and thermal management applications is their surface roughness. Growth of the diamond phase on a nondiamond substrate is initiated by nucleation at either randomly seeded sites or at thermally favored sites, due to statistical thermal fluctuations at the substrate surface. Depending on the growth temperature and pressure conditions, certain favored crystal orientations dominate the competitive growth process. As a result, the films grown are polycrystalline in nature with a relatively large grain size ($> 1\,\mu m$) and they terminate in very rough surfaces, with RMS roughnesses ranging from a few tenths of a micrometer to tens of micrometers. Techniques for polishing these films have been developed. It has been reported that laser polished films exhibit friction and wear properties almost comparable to those of bulk polished diamond [38.7, 8].

Amorphous carbon has no long-range order, and the short-range order of the carbon atoms in it can have one or more of three bonding configurations: sp^3 (diamond), sp^2 (graphite), or sp^1 (with two electrons forming strong σ bonds, and the remaining two electrons left in orthogonal p_y and p_z orbitals, that form weak π bonds). Short-range order controls the properties of amorphous

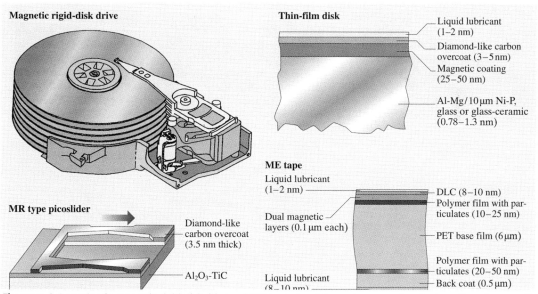

Fig. 38.2 Schematic of a magnetic rigid-disk drive and MR type picoslider, and cross-sectional schematics of a magnetic thin film rigid disk and a metal evaporated (ME) tape

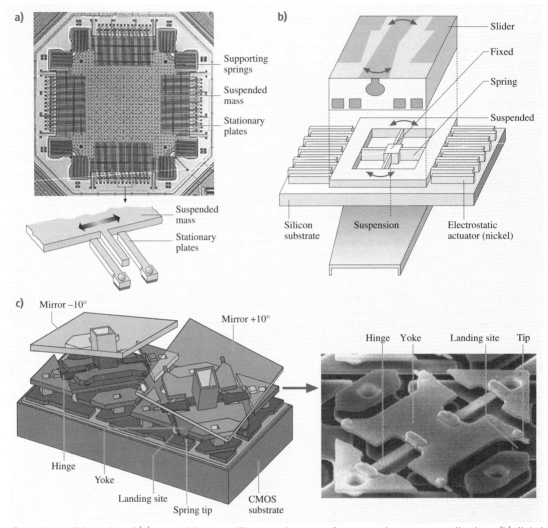

Fig. 38.3a–c Schematics of (**a**) a capacitive-type silicon accelerometer for automotive sensory applications, (**b**) digital micrometer devices for high-projection displays, and (**c**) a polysilicon rotary microactuator for a magnetic disk drives

materials and coatings. Hard amorphous carbon (a-C) coatings, commonly known as diamond-like carbon or DLC (implying high hardness) coatings, are a class of coatings that are mostly metastable amorphous materials, but that include a micro- or nanocrystalline phase. The coatings are random networks of covalently bonded carbon in hybridized tetragonal (sp^3) and trigonal (sp^2) local coordination with some of the bonds terminated by hydrogen. These coatings have been successfully deposited by a variety of vacuum deposition techniques on a variety of substrates at or near room temperature. These coatings generally reproduce substrate topography and do not require any post-finishing. However, these coatings mostly adhere best on silicon substrates. The best adhesion is obtained on substrates that form carbides, such as Si, Fe and Ti. Based on depth profile analyses (using Auger and XPS) of DLC coatings deposited on silicon substrates, it has been reported that a substantial amount of silicon carbide (on the order of 5–10 nm in thickness) is present at the carbon–silicon interface, giving good adhesion and hardness [38.9]. For good adhesion of DLC coatings to other substrates, in most cases, an interlayer of silicon is required in most cases, except for coatings deposited by a cathodic arc.

There is significant interest in DLC coatings due to their unique combination of desirable properties. These properties include high hardness and wear resistance, chemical inertness to both acids and alkalis, lack of magnetic response, and an optical band gap ranging from zero to a few eV, depending upon the deposition conditions. These are used in a wide range of applications, including tribological, optical, electronic and biomedical applications [38.1, 10, 11]. The high hardness, good friction and wear properties, versatility in deposition and substrates, and no requirement for post-deposition finishing make them very attractive for tribological applications. Two primary examples include overcoats for magnetic media (thin film disks and ME tapes) and MR-type magnetic heads for magnetic storage devices (Fig. 38.2) [38.12–20], and the emerging field of microelectromechanical systems (Fig. 38.3) [38.21–24]. The largest industrial application of the family of amorphous carbon coatings, typically deposited by DC/RF magnetron sputtering, plasma-enhanced chemical vapor deposition or ion beam deposition techniques, is in magnetic storage devices. These are employed to protect magnetic coatings on thin film rigid disks and metal evaporated tapes and the thin film head structure of a read/write disk head against wear and corrosion (Fig. 38.2). Thicknesses ranging from 3 to 10 nm are employed to maintain low physical spacing between the magnetic element of a read/write head and the magnetic layer of the storage media. Mechanical properties affect friction wear and therefore need to be optimized. In 1998, Gillette introduced Mach 3 razor blades with ultrathin DLC coatings, which could potentially become a very large industrial application. DLC coatings are also used in other commercial applications such as the glass windows of supermarket laser barcode scanners and sunglasses. These coatings are actively pursued in microelectromechanical systems (MEMS) components [38.23].

In this chapter, a state-of-the-art review of recent developments in the field of chemical, mechanical, and tribological characterization of ultrathin amorphous carbon coatings is presented. An overview of the most commonly used deposition techniques is provided, followed by typical chemical and mechanical characterization data and typical tribological data from both coupon-level testing and functional testing.

38.1 Description of Common Deposition Techniques

The first hard amorphous carbon coatings were deposited by a beam of carbon ions produced in an argon plasma on room temperature substrates, as reported by *Aisenberg* and *Chabot* [38.25]. Subsequent confirmation by *Spencer* et al. [38.26] led to the explosive growth of this field. Following this first work, several alternative techniques were developed. Amorphous carbon coatings have been prepared by a variety of deposition techniques and precursors, including evaporation, DC, RF or ion beam sputtering, RF or DC plasma-enhanced chemical vapor deposition (PECVD), electron cyclotron resonance chemical vapor deposition (ECR-CVD), direct ion beam deposition, pulsed laser vaporization and vacuum arc, from a variety of carbon-bearing solids or gaseous source materials [38.1, 27]. Coatings with both graphitic and diamond-like properties have been produced. Evaporation and ion plating techniques have been used to produce coatings with graphitic properties (low hardness, high electrical conductivity, very low friction, and so on, and all of the techniques have been used to produce coatings with diamond-like properties.

The structure and properties of a coating are dependent upon the deposition technique and parameters. High-energy surface bombardment has been used to produce harder and denser coatings. It is reported that the sp^3/sp^2 fraction decreases in the order: cathodic arc deposition, pulsed laser vaporization, direct ion beam deposition, plasma-enhanced chemical vapor deposition, ion beam sputtering, DC/RF sputtering [38.12, 28, 29]. A common feature of these techniques is that the deposition is energetic; in other words the carbon species arrive with an energy significantly greater than that represented by the substrate temperature. The resultant coatings are amorphous in structure, with hydrogen contents of up to 50%, and display a high degree of sp^3 character. From the results of previous investigations, it has been proposed that deposition of sp^3-bonded carbon requires that the depositing species have kinetic energies on the order of 100 eV or higher, well above those obtained in thermal processes like evaporation (0–0.1 eV). The species must then be quenched into the metastable configuration via rapid energy removal. Ex-

Table 38.1 Summary of common deposition techniques, the kinetic energies of the depositing species and deposition rates

Deposition technique	Process	Kinetic energy (eV)	Deposition rate (nm/s)
Filtered cathodic arc (FCA)	Energetic carbon ions produced by a vacuum arc discharge between a graphite cathode and a grounded anode	100–2500	0.1–1
Direct ion beam (IB)	Carbon ions produced from methane gas in an ion source and accelerated toward a substrate	50–500	0.1–1
Plasma-enhanced chemical vapor deposition (PECVD)	Hydrocarbon species produced by plasma decomposition of a hydrocarbon gas (such as acetylene) are accelerated toward a DC-biased substrate	1–30	1–10
Electron cyclotron resonance plasma chemical vapor deposition (ECR-CVD)	Hydrocarbon ions produced by the plasma decomposition of ethylene gas in the presence of a plasma at the electron cyclotron resonance condition are accelerated toward a RF-biased substrate	1–50	1–10
DC/RF sputtering	Sputtering of graphite target by argon ion plasma	1–10	1–10

cess energy, such as that provided by substrate heating, is detrimental to the achievement of a high sp^3 fraction. In general, the higher the fraction of sp^3-bonded carbon atoms in the amorphous network, the greater the hardness [38.29–36]. The mechanical and tribological properties of a carbon coating depend on the sp^3/sp^2-bonded carbon ratio, the amount of hydrogen in the coating, and the adhesion of the coating to the substrate, which are influenced by the precursor material, the kinetic energy of the carbon species prior to deposition, the deposition rate, the substrate temperature, the substrate biasing, and the substrate itself [38.29, 33, 35, 37–46]. The kinetic energies and deposition rates involved in selected deposition processes used in the deposition of DLC coatings are compared in Table 38.1 [38.1, 28].

In the studies by *Gupta* and *Bhushan* [38.12, 47], *Li* and *Bhushan* [38.48, 49], and *Sundararajan* and *Bhushan* [38.50], DLC coatings typically ranging in thickness from 3.5 nm to 20 nm were deposited on single-crystal silicon, magnetic Ni-Zn ferrite, and Al_2O_3-TiC substrates (surface roughness ≈ 1–3 nm RMS) by filtered cathodic arc (FCA) deposition, (direct) ion beam deposition (IBD), electron cyclotron resonance chemical vapor deposition (ECR-CVD), plasma-enhanced chemical vapor deposition (PECVD), and DC/RF planar magnetron sputtering (SP) deposition techniques [38.51]. In this chapter, we will limit the presentation of data to coatings deposited by FCA, IBD, ECR-CVD and SP deposition techniques.

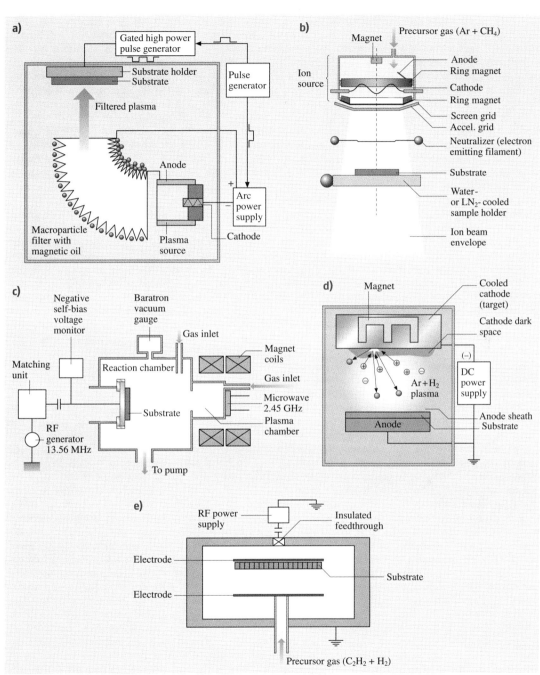

Fig. 38.4a–e Schematic diagrams of deposition by (**a**) filtered cathodic arc deposition, (**b**) ion beam deposition, (**c**) electron cyclotron resonance chemical vapor deposition (ECR-CVD), (**d**) DC planar magnetron sputtering, and (**e**) plasma-enhanced chemical vapor deposition (PECVD)

38.1.1 Filtered Cathodic Arc Deposition

When the filtered cathodic arc deposition technique is used to create carbon coatings [38.29, 52–59], a vacuum arc plasma source is used to form the carbon film. In the FCA technique used by *Gupta* and *Bhushan* [38.12], energetic carbon ions are produced by a vacuum arc discharge between a planar graphite cathode and a grounded anode (Fig. 38.4a). The cathode is a 6 mm diameter high-density graphite disk mounted on a water-cooled copper block. The arc is driven at an arc current of 200 A, with an arc duration of 5 ms and an arc repetition rate of 1 Hz. The plasma beam is guided by a magnetic field that transports current between the electrodes to form tiny, rapidly moving spots on the cathode surface. The source is coupled to a 90° bent magnetic filter to remove the macroparticles produced concurrently with the plasma in the cathode spots. The ion current density at the substrate is in the range of 10–50 mA/cm^2. The base pressure is less than 10^{-4} Pa. A much higher plasma density is achieved using a powerful arc discharge than using electron beam evaporation with auxiliary discharge. In the discharge process, the cathodic material suffers a complicated transition from the solid phase to an expanding, nonequilibrium plasma via liquid and dense equilibrium nonideal plasma phases [38.58]. The carbon ions in the vacuum arc plasma have a direct kinetic energy of 20–30 eV. The high voltage pulses are applied to the substrate which is mounted on a water-cooled sample holder, and ions are accelerated through the sheath and arrive at the substrate with an additional energy given by the potential difference between the plasma and the substrate. The substrate holder is pulsed-biased to a voltage of up to -2 kV with a pulse duration of 1 μs. The negative biasing of -2 kV corresponds to a kinetic energy of 2 keV for the carbon ions. The use of a pulsed bias instead of a DC bias enables much higher voltages to be applied and it permits a surface potential to be created on nonconducting films. The ion energy is varied during the deposition. For the first 10% of the deposition, the substrates are pulsed-biased to -2 keV with a pulse duty cycle of 25%, so for 25% of the time the energy is 2 keV and for the remaining 75% it is 20 eV, which is the *natural* energy of carbon ions in a vacuum discharge. For the last 90% of the deposition, the pulsed-biased voltage is reduced to -200 eV with a pulsed bias duty cycle of 25%, so the energy is 200 eV for 25% and 20 eV for 75% of the deposition. The high energy at the beginning leads to good intermixing and adhesion of the films, whereas the lower energy at the later stage leads to hard films. Under the conditions described, the deposition rate at the substrate is about 0.1 nm/s, which is slow. Compared with most gaseous plasma, the cathodic arc plasma is almost fully ionized, and the ionized carbon atoms have high kinetic energies which promotes the formation of a high fraction of sp^3-bonded carbon ions, which in turn results in high hardness and higher interfacial adhesion. *Cuomo* et al. [38.42] have reported, based on electron energy loss spectroscopy (EELS) analysis, that the sp^3-bonded carbon fraction of a cathodic arc coating is 83% compared to 38% for ion beam sputtered carbon. These coatings are reported to be *nonhydrogenated*.

This technique does not require an adhesion underlayer for nonsilicon substrates. However, adhesion of the DLC coatings on the electrically insulating substrate is poor, as negative pulsed biasing forms an electrical sheath that accelerates depositing ions to the substrate and enhances the adhesion of the coating to the substrate with associated ion implantation. It is difficult to build potential on an insulating substrate, and lack of biasing results in poor adhesion.

38.1.2 Ion Beam Deposition

In the direct ion beam deposition of a carbon coating [38.60–64], as used by *Gupta* and *Bhushan* [38.12], the carbon coating is deposited from an accelerated carbon ion beam. The sample is precleaned by ion etching. In the case of nonsilicon substrates, a 2–3 nm thick amorphous silicon adhesion layer is deposited by ion beam sputtering using an ion beam containing a mixture of methane and argon at 200 V. For the carbon deposition, the chamber is pumped to about 10^{-4} Pa, and methane gas is fed through the cylindrical ion source and ionized by energetic electrons produced by a hot-wire filament (Fig. 38.4b). Ionized species then pass through a grid with a bias voltage of about 50 eV, where they gain a high acceleration energy and reach a hot-wire filament, emitting thermionic electrons that neutralize the incoming ions. The discharging of ions is important when insulating ceramics are used as substrates. The species are then deposited on a water-cooled substrate. Operating conditions are adjusted to give an ion beam with an acceleration energy of about 200 eV and a current density of about 1 mA/cm^2. At these operating conditions, the deposition rate is about 0.1 nm/s, which is slow. Incidentally, tough and soft coatings are deposited at a high acceleration energy of about 400 eV and at a deposition rate of about 1 nm/s. The ion beam-deposited carbon coatings are reported to be hydrogenated (30–40 at. % hydrogen).

38.1.3 Electron Cyclotron Resonance Chemical Vapor Deposition

The lack of electrodes in the ECR-CVD technique and its ability to create high densities of charged and excited species at low pressures ($\leq 10^{-4}$ Torr) make it attractive for coating deposition [38.65]. In the ECR-CVD carbon deposition process described by *Suzuki* and *Okada* [38.66] and used by *Li* and *Bhushan* [38.48, 49] and *Sundararajan* and *Bhushan* [38.50], microwave power is generated by a magnetron operating in continuous mode at a frequency of 2.45 GHz (Fig. 38.4c). The plasma chamber functions as a microwave cavity resonator. The magnetic coils arranged around the plasma chamber generate a magnetic field of 875 G, necessary for electron cyclotron resonance. The substrate is placed on a stage that is connected capacitively to a 13.56 MHz RF generator. The process gas is introduced into the plasma chamber and the hydrocarbon ions generated are accelerated by a negative self-bias voltage, which is generated by applying RF power to the substrate. Both the substrate stage and the plasma chamber are water-cooled. The process gas used is 100% ethylene and its flow rate is held constant at 100 sccm. The microwave power is 100–900 W. The RF power is 30–120 W. The pressure during deposition is kept close to the optimum value of 5.5×10^{-3} Torr. Before the deposition, the substrates are cleaned using Ar ions generated in the ECR plasma chamber.

38.1.4 Sputtering Deposition

In DC planar magnetron carbon sputtering [38.13, 33, 37, 40, 67–71], the carbon coating is deposited by the sputtering of a graphite target with Ar ion plasma. In the glow discharge, positive ions from the plasma strike the target with sufficient energy to dislodge the atoms by momentum transfer, which are intercepted by the substrate. An ≈ 5 nm thick amorphous silicon adhesion layer is initially deposited by sputtering if the deposition is to be carried out on a nonsilicon surface. In the process used by *Gupta* and *Bhushan* [38.12], the coating is deposited by the sputtering of a 200 mm diameter graphite target with Ar ion plasma at 300 W power and a pressure of about 0.5 Pa (6 mTorr) (Fig. 38.4d). Plasma is generated by applying a DC potential between the substrate and a target. *Bhushan* et al. [38.35] reported that the sputtered carbon coating contains about 35 at. % hydrogen. The hydrogen comes from the hydrocarbon contaminants present in the deposition chamber. In order to produce a hydrogenated carbon coating with a larger concentration of hydrogen, the deposition is carried out in Ar and hydrogen plasma.

38.1.5 Plasma-Enhanced Chemical Vapor Deposition

In the RF-PECVD deposition of carbon, as used by *Gupta* and *Bhushan* [38.12], the carbon coating is deposited by adsorbing hydrocarbon free radicals onto the substrate and then via chemical bonding to other atoms on the surface. The hydrocarbon species are produced by the RF plasma decomposition of hydrocarbon precursors such as acetylene (C_2H_2), Fig. 38.4e [38.27, 69, 72–75]. Instead of requiring thermal energy, as in thermal CVD, the energetic electrons in the plasma (at a pressure of $1-5 \times 10^2$ Pa, and typically less than 10 Pa) can activate almost any reaction among the gases in the glow discharge at relatively a low substrate temperature of 100 to 600 °C (typically less than 300 °C). To deposit the coating on nonsilicon substrates, an amorphous silicon adhesion layer about 4 nm thick is first deposited under similar conditions from a gas mixture of 1% silane in argon in order to improve adhesion [38.76]. In the process used by *Gupta* and *Bhushan* [38.12], the plasma is sustained in a parallel-plate geometry by a capacitive discharge at 13.56 MHz, at a surface power density of around 100 mW/cm^2. The deposition is performed at a flow rate on the order of 6 sccm and a pressure on the order of 4 Pa (30 mTorr) on a cathode-mounted substrate maintained at a substrate temperature of 180 °C. The cathode bias is held fixed at about -120 V with an external DC power supply attached to the substrate (powered electrode). The carbon coatings deposited by PECVD usually contain hydrogen at levels of up to 50% [38.35, 77].

38.2 Chemical and Physical Coating Characterization

The chemical structures and properties of amorphous carbon coatings depend on the deposition conditions employed when they are formed. It is important to understand the relationship between the chemical structure of a coating and its properties since it allows useful deposition parameters to be defined. Amorphous carbon films are metastable phases formed when carbon particles are condensed on a substrate. The prevailing atomic arrangement in the DLC coatings is amorphous or quasi-amorphous, with small diamond (sp^3), graphite (sp^2) and other unidentifiable micro- or nanocrystallites. The coating is dependent upon the deposition process and the deposition conditions used because these influence the sp^3/sp^2 ratio and the proportion of hydrogen in the coating. The sp^3/sp^2 ratios of DLC coatings typically range from 50% to close to 100%, and hardness increases with the sp^3/sp^2 ratio. Most DLC coatings, except those produced by a filtered cathodic arc, contain from a few to about 50 at. % hydrogen. Sometimes hydrogen and nitrogen are deliberately added to produce hydrogenated (a-C:H) and nitrogenated amorphous carbon (a-C:N) coatings, respectively. Hydrogen helps to stabilize sp^3 sites (most of the carbon atoms attached to hydrogen have a tetrahedral structure), so the sp^3/sp^2 ratio for hydrogenated carbon is higher [38.30]. The optimum sp^3/sp^2 ratio for a random covalent network composed of sp^3 and sp^2 carbon sites (N_{sp^2} and N_{sp^3}) and hydrogen is [38.30]

$$\frac{N_{sp^3}}{N_{sp^2}} = \frac{6X_H - 1}{8 - 13X_H}, \quad (38.1)$$

where X_H is the atomic fraction of hydrogen. The hydrogenated carbon has a larger optical band gap, higher electrical resistivity (semiconductor), and a lower optical absorption or high optical transmission. Hydrogenated coatings have lower densities, probably because of the reduced cross-linking due to hydrogen incorporation. However, the hardness decreases with increasing hydrogen, even though the proportion of sp^3 sites increases (that is, as the local bonding environment becomes more diamondlike) [38.78, 79]. It is speculated that the high hydrogen content introduces frequent terminations in the otherwise strong 3-D network, and hydrogen increases the soft polymeric component of the structure more than it enhances the cross-linking sp^3 fraction.

A number of investigations have been performed to identify the microstructure of amorphous carbon films using a variety of techniques, such as Raman spectroscopy, EELS, nuclear magnetic resonance, optical measurements, transmission electron microscopy, and x-ray photoelectron spectroscopy [38.33]. The structure of diamondlike amorphous carbon is amorphous or quasi-amorphous, with small graphitic (sp^2), tetrahedrally coordinated (sp^3) and other types of nanocrystallites (typically on the order of a couple of nm in size, randomly oriented) [38.33, 80, 81]. These studies indicate that the chemical structure and physical properties of the coatings are quite variable, depending on the deposition techniques and film growth conditions. It is clear that both sp^2- and sp^3-bonded atomic sites are incorporated in diamondlike amorphous carbon coatings and that the physical and chemical properties of the coatings depend strongly on their chemical bonding and microstructures. Systematic studies have been conducted to carry out chemical characterization and to investigate how the physical and chemical properties of amorphous carbon coatings vary as a function of deposition conditions [38.33, 35, 40]. EELS and Raman spectroscopy are commonly used to characterize the chemical bonding and microstructure. The hydrogen concentration in the coating is obtained via forward recoil spectrometry (FRS). A variety of physical properties relevant to tribological performance are measured.

In order to give the reader a feel for typical data obtained when characterizing amorphous carbon coatings and their relationships to physical properties, we present data on several sputtered coatings, RF-PECVD amorphous carbon and microwave-PECVD (MPECVD) diamond coatings [38.33, 35, 40]. The sputtered coatings were DC magnetron sputtered at a chamber pressure of 10 mTorr under sputtering power densities of 0.1 and 2.1 W/cm^2 (labeled as coatings W1 and W2, respectively) in a pure Ar plasma. These coatings were prepared at a power density of 2.1 W/cm^2 with various hydrogen fractions (0.5, 1, 3, 5, 7 and 10%) of Ar/H; the gas mixtures were labeled as H1, H2, H3, H4, H5, and H6, respectively.

38.2.1 EELS and Raman Spectroscopy

EELS and Raman spectra of four sputtered (W1, W2, H1, and H3) carbon samples and one PECVD carbon sample were obtained. Figure 38.5 shows the EELS spectra of these carbon coatings. EELS spectra (up to 50 eV) for bulk diamond and polycrystalline graphite are also shown in Fig. 38.5. One prominent peak is seen at 35 eV in diamond, while two peaks are seen at 27 eV and 6.5 eV in graphite, which are called the ($\pi + \sigma$) and (π) peaks, respectively. These peaks are produced by the loss of transmitted electron energy to plasmon oscillations of the valence electrons. The $\pi + \sigma$ peak in each coating is positioned at a lower energy region than that of graphite. The π peaks in the W series and PECVD samples also occur at a lower energy region than that of the graphite. However, the π

peaks in the H-series are comparable to or higher than those of graphite (Table 38.2). The plasmon oscillation frequency is proportional to the square root of the corresponding electron density to a first approximation. Therefore, the samples in the H-series most likely have a higher density of π electrons than the other samples.

Amorphous carbon coatings contain (mainly) a mixture of sp^2- and sp^3-bonds, even though there is some evidence for the presence of sp-bonds as well [38.82]. The PECVD coatings and the H-series coatings in this study have almost the same mass density (as seen in Table 38.4, discussed in more detail later), but the former have a lower concentration of hydrogen (18.1%) than the H-series (35–39%) (as seen in Table 38.3, also discussed in more detail later). The relatively low-energy positions of the π peaks of the PECVD coatings compared to those of the H-series indicates that the PECVD coatings contain a higher fraction of sp^3-bonds than the sputtered hydrogenated carbon coatings (H-series).

Figure 38.5b shows EELS spectra associated with the inner-shell (K-shell) ionization. Again, the spectra for diamond and polycrystalline graphite are included for comparison. Sharp peaks are observed at 285.5 eV and 292.5 eV in graphite, while no peak is seen at 285.5 eV in diamond. The general features of the K-shell EELS spectra for the sputtered and PECVD carbon samples resemble those of graphite, but with the higher energy features smeared. The peak at 285.5 eV in the sputtered and PECVD coatings also indicates the presence of sp^2-bonded atomic sites in the coatings. All of these spectra peak at 292.5 eV, similar to the spectra of graphite, but the peak in graphite is sharper.

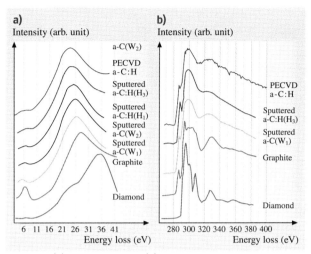

Fig. 38.5 (a) Low-energy and (b) high-energy EELS of DLC coatings produced by the DC magnetron sputtering and RF-PECVD techniques. Data for bulk diamond and polycrystalline graphite are included for comparison [38.35]

Table 38.2 Experimental results from EELS and Raman spectroscopy [38.35]

Sample	EELS peak position		Raman peak position		Raman FWHM[a]		I_D/I_G^d
	π (eV)	$\pi + \sigma$ (eV)	G-band[b] (cm^{-1})	D-band[c] (cm^{-1})	G-band (cm^{-1})	D-band (cm^{-1})	
Sputtered a-C coating (W1)	5.0	24.6	1541	1368	105	254	2.0
Sputtered a-C coating (W2)	6.1	24.7	1560	1379	147	394	5.3
Sputtered a-C:H coating (H1)	6.3	23.3	1542	1334	95	187	1.6
Sputtered a-C:H coating (H3)	6.7	22.4	e	e	e	e	e
PECVD a-C:H coating	5.8	24.0	1533	1341	157	427	1.5
Diamond coating	–	–	1525[f]	1333[g]	–	8[g]	–
Graphite (for reference)	6.4	27.0	1580	1358	37	47	0.7
Diamond (for reference)	–	37.0	–	1332[g]	–	2[g]	–

[a] Full width at half maximum
[b] Peak associated with sp^2 *graphite* carbon
[c] Peak associated with sp^2 *disordered* carbon (not sp^3-bonded carbon)
[d] Intensity ratio of the D-band to the G-band
[e] Fluorescence
[f] Includes D- and G-band, signal too weak to analyze
[g] Peak position and width for diamond phonon

Table 38.3 Experimental results of FRS analysis [38.35]

Sample	Ar/H ratio	C (at.% ± 0.5)	H (at.% ± 0.5)	Ar (at.% ± 0.5)	O (at.% ± 0.5)
Sputtered a-C coating (W2)	100/0	90.5	9.3	0.2	–
Sputtered a-C:H coating (H2)	99/1	63.9	35.5	0.6	–
Sputtered a-C:H coating (H3)	97/3	56.1	36.5	–	7.4
Sputtered a-C:H coating (H4)	95/5	53.4	39.4	–	7.2
Sputtered a-C:H coating (H5)	93/7	58.2	35.4	0.2	6.2
Sputtered a-C:H coating (H6)	90/10	57.3	35.5	–	7.2
PECVD a-C:H coating	99.5% CH_4	81.9	18.1	–	–
Diamond coating	H_2-1 mol % CH_4	94.0	6.0	–	–

Table 38.4 Experimental results of physical properties [38.35]

Sample	Mass density (g/cm^3)	Nano-hardness (GPa)	Elastic modulus (GPa)	Electrical resistivity (Ω cm)	Compressive residual stress (GPa)
Sputtered a-C coating (W1)	2.1	15	141	1300	0.55
Sputtered a-C:H coating (W2)	1.8	14	136	0.61	0.57
Sputtered a-C:H coating (H1)	–	14	96	–	> 2
Sputtered a-C:H coating (H3)	1.7	7	35	> 10^6	0.3
PECVD a-C:H coating	1.6–1.8	33–35	≈ 200	> 10^6	1.5–3.0
Diamond coating	–	40–75	370–430	–	–
Graphite (for reference)	2.267	Soft	9–15	5×10^{-5a}, 4×10^{-3b}	0
Diamond (for reference)	3.515	70–102	900–1050	10^7–10^{20}	0

^{a} Parallel to layer planes
^{b} Perpendicular to layer planes

Raman spectra from samples W1, W2, H1 and PECVD are shown in Fig. 38.6. Raman spectra could not be observed in specimens H2 and H3 due to high fluorescence signals. The Raman spectra of single-crystal diamond and polycrystalline graphite are also shown for comparison in Fig. 38.6. The results from the spectral fits are summarized in Table 38.2. We will focus on the position of the G-band, which has been shown to be related to the fraction of sp^3-bonded sites. Increasing the power density in the amorphous carbon coatings (W1 and W2) results in a higher G-band frequency, implying a smaller fraction of sp^3-bonding in W2 than in W1. This is consistent with the higher density of W1. H1 and PEVCD have even lower G-band positions than W1, implying an even higher fraction of sp^3-bonding, which is presumably caused by the incorporation of H atoms into the lattice. The high hardness of H3 might be attributed to efficient sp^3 cross-linking of small sp^2-ordered domains.

The Raman spectrum of a MPECVD diamond coating is shown in Fig. 38.6. The Raman peak of diamond is at 1333 cm^{-1}, with a line width of 7.9 cm^{-1}. There is a small broad peak at around 1525 cm^{-1}, which is attributed to a small amount of a-C:H. This impurity peak is not intense enough to be able to separate the G- and D-bands. The diamond peak frequency is very close to that of natural diamond (1332.5 cm^{-1}, see Fig. 38.6), indicating that the coating is not under stress [38.83]. The large line width compared to that of natural diamond (2 cm^{-1}) indicates that the microcrystallites probably have a high concentration of defects [38.84].

38.2.2 Hydrogen Concentrations

A FRS analysis of six sputtered (W2, H2, H3, H4, H5, and H6) coatings, one PECVD coating, and one diamond coating was performed. Figure 38.7 shows an overlay of the spectra from the six sputtered samples. Similar spectra were obtained from the PECVD and the

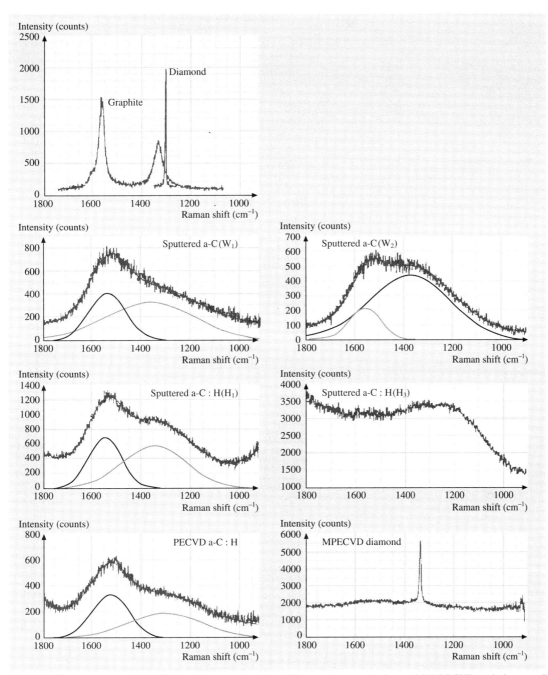

Fig. 38.6 Raman spectra of the DLC coatings produced by DC magnetron sputtering and RF-PECVD techniques and a diamond film produced by the MPE-CVD technique. Data for bulk diamond and microcrystalline graphite are included for comparison [38.35]

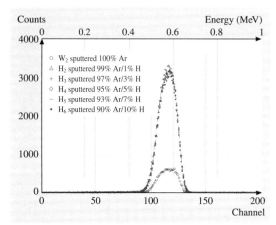

Fig. 38.7 FRS spectra for six DLC coatings produced by DC magnetron sputtering [38.35]

diamond films. Table 38.3 shows the H and C fractions as well as the amount of impurities (Ar and O) in the films in atomic %. Most apparent is the large fraction of H in the sputtered films. Regardless of how much H_2 is in the Ar sputtering gas, the H content of the coatings is about the same, ≈ 35 at. %. Interestingly, there is still ≈ 10% H present in the coating sputtered in pure Ar (W2). It is interesting to note that Ar is present only in coatings grown using Ar carrier gas with a low (< 1%) H content. The presence of O in the coatings combined with the fact that the coatings were prepared approximately nine months before the FRS analysis caused suspicion that they had absorbed water vapor, and that this may be the cause of the H peak in specimen W2.

All samples were annealed for 24 h at 250 °C in a flowing He furnace and then reanalyzed. Surprisingly, the H contents of all coatings measured increased slightly, even though the O content decreased, and W2 still had a substantial amount of H_2. This slight increase in H concentration is not understood. However, the fact that the H concentration did not decrease with the oxygen as a result of annealing suggests that high H concentration is not due to adsorbed water vapor. The PECVD film has more H (≈ 18%) than the sputtered films initially, but after annealing it has the same fraction as specimen W2, the film sputtered in pure Ar. The diamond film has the smallest amount of hydrogen, as seen in Table 38.3.

38.2.3 Physical Properties

The physical properties of the four sputtered (W1, W2, H1, and H3) coatings, one PECVD coating, one diamond coating, and bulk diamond and graphite are presented in Table 38.4. The hydrogenated carbon and the diamond coatings have very high resistivity compared to unhydrogenated carbon coatings. It appears that unhydrogenated carbon coatings have higher densities than hydrogenated carbon coatings, although both groups are less dense than graphite. The density depends upon the deposition technique and the deposition parameters. It appears that unhydrogenated sputtered coatings deposited at low power exhibit the highest density. The nanohardness of hydrogenated carbon is somewhat lower than that of unhydrogenated carbon. PECVD coatings are significantly harder than sputtered coatings. The nanohardness and modulus of elasticity of a diamond coating is very high compared to that of a DLC coating, even though the hydrogen contents are similar. The compressive residual stresses of the PECVD coatings are substantially higher than those of sputtered coatings, which is consistent with the results for the hardness.

Figure 38.8a shows the effect of hydrogen in the plasma on the residual stresses and the nanohardness for the sputtered coatings W2 and H1 to H6. The coat-

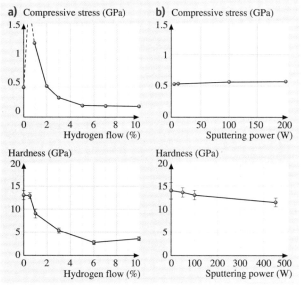

Fig. 38.8a,b Residual compressive stresses and nanohardness (**a**) as a function of hydrogen flow rate, where the sputtering power is 100 W and the target diameter is 75 mm (power density = 2.1 W/cm^2), and (**b**) as a function of sputtering power over a 75 mm diameter target with no hydrogen added to the plasma [38.40]

ings made with a hydrogen flow of between 0.5 and 1.0% delaminate very quickly, even when they are only a few tens of nm thick. In pure Ar and at H_2 flows that are greater than 1%, the coatings appear to be more adhesive. The tendency of some coatings to delaminate can be caused by intrinsic stress in the coating, which is measured by substrate bending. All of the coatings in the figure are in compressive stress. The maximum stress occurs between 0 and 1% H_2 flow, but the stress cannot be quantified in this range because the coatings instantly delaminate upon exposure to air. At higher hydrogen concentrations the stress gradually diminishes. A generally decreasing trend is observed for the hardness of the coatings as the hydrogen content is increased. The hardness decreases slightly, going from 0% H_2 to 0.5% H_2, and then decreases sharply. These results are probably lower than the true values because of local delamination around the indentation point. This is especially likely for the 0.5% and 1.0% coatings, where delamination is visually apparent, but may also be true to a lesser extent for the other coatings. Such an adjustment would bring the hardness profile into closer correlation with the stress profile. *Weissmantel* et al. [38.68] and *Scharff* et al. [38.85] observed a downturn in hardness for high bias and a low hydrocarbon gas pressure for ion-plated carbon coating, and, therefore, presumably low hydrogen content in support of the above contention.

Figure 38.8b shows the effect of sputtering power (with no hydrogen added to the plasma) on the residual stresses and nanohardness for various sputtered coatings. As the power decreases, the compressive stress does not seem to change while the nanohardness slowly increases. The rate of change becomes more rapid at very low power levels.

The addition of H_2 during sputtering of the carbon coatings increases the H concentration in the coating. Hydrogen causes the character of the C–C bonds to shift from sp^2 to sp^3, and a rise in the number of C–H bonds, which ultimately relieves stress and produces a softer *polymerlike* material. Low power deposition, like the presence of hydrogen, appears to stabilize the formation of sp^3 C–C bonds, increasing hardness. These coatings relieve stress and lead to better adhesion. Increasing the temperature during deposition at high power density results in graphitization of the coating material, producing a decrease in hardness with an increase in power density. Unfortunately, low power also means impractically low deposition rates.

38.2.4 Summary

Based on analyses of EELS and Raman data, it is clear that all DLC coatings have both sp^2 and sp^3 bonding characteristics. The sp^2/sp^3 bonding ratio depends upon the deposition technique and parameters used. DLC coatings deposited by sputtering and PECVD contain significant concentrations of hydrogen, while diamond coatings contain only small amounts of hydrogen impurities. Sputtered coatings with no deliberate addition of hydrogen in the plasma contain a significant amount of hydrogen. Regardless of how much hydrogen is in the Ar sputtering gas, the hydrogen content of the coatings increases initially but then does not increase further.

Hydrogen flow and sputtering power density affect the mechanical properties of these coatings. Maximum compressive residual stress and hardness occur between 0 and 1% hydrogen flow, resulting in rapid delamination. Low sputtering power moderately increases hardness and also relieves residual stress.

38.3 Micromechanical and Tribological Coating Characterization

38.3.1 Micromechanical Characterization

Common mechanical characterizations include measurements of hardness and elastic modulus, fracture toughness, fatigue life, and scratch and wear testing. Nanoindentation and atomic force microscopy (AFM) are used for the mechanical characterization of ultrathin films.

Hardness and elastic modulus are calculated from the load displacement data obtained by nanoindentation at loads of typically 0.2 to 10 mN using a commercially available nanoindenter [38.23, 86]. This instrument monitors and records the dynamic load and displacement of the three-sided pyramidal diamond (Berkovich) indenter during indentation. For fracture toughness measurements of ultrathin films 100 nm to a few μm thick, a nanoindentation-based technique is used in which through-thickness cracking in the coating is detected from a discontinuity observed in the load–displacement curve, and the energy released during cracking is obtained from the curve [38.87–89]. Based on the energy released, fracture mechanics analysis is then used to calculate the fracture toughness. An indenter with a cube-corner tip geometry is preferred because the through-thickness cracking of hard films can be accomplished at lower loads. In fatigue measurement, a conical diamond indenter with a tip radius of about 1 μm is used and load cycles with sinusoidal shapes are applied [38.90, 91]. The fatigue behavior of a coating is studied by monitoring the change in contact stiffness, which is sensitive to damage formation.

Hardness and Elastic Modulus

For materials that undergo plastic deformation, high hardness and elastic modulus are generally needed for low friction and wear, whereas for brittle materials, high fracture toughness is needed [38.2, 3, 21]. The DLC coatings used for many applications are hard and brittle, and values of hardness and fracture toughness need to be optimized.

Representative load–displacement plots of indentations made at 0.2 mN peak indentation load on 100 nm thick DLC coatings deposited by the four deposition techniques on a single-crystal silicon substrate are compared in Fig. 38.9. The indentation depths at the peak load range from about 18 to 26 nm, smaller than that of the coating thickness. Many of the coatings exhibit a discontinuity or pop-in marks in the loading curve, which indicate a sudden penetration of the tip into the sample. A nonuniform penetration of the tip into a thin coating possibly results from formation of cracks in the coating, formation of cracks at the coating–substrate interface, or debonding or delamination of the coating from the substrate.

The hardness and elastic modulus values for a peak load of 0.2 mN on the various coatings and single-crystal silicon substrate are summarized in Table 38.5 and Fig. 38.10 [38.47, 49, 89, 90]. Typical values for the peak and residual indentation depths range from 18 to 26 nm and 6 to 12 nm, respectively. The FCA coating exhibits the greatest hardness of 24 GPa and the highest elastic modulus of 280 GPa of the various coatings, followed by the ECR-CVD, IB and SP coatings. Hardness and elastic modulus have been known to vary over a wide range with the sp^3-to-sp^2 bonding ratio, which depends on the kinetic energy of the carbon species and the amount of hydrogen [38.6, 30, 47, 92, 93]. The high hardness and elastic modulus of the FCA coatings are attributed to the high kinetic energy of the carbon species involved in the FCA deposition [38.12, 47]. *Anders* et al. [38.57] also reported a high hardness, measured by nanoindentation, of about 45 GPa for cathodic arc carbon coatings. They observed a change in hardness from 25 to 45 GPa with a pulsed bias voltage and bias duty cycle. The high hardness of cathodic arc carbon was attributed to the high percentage (more than 50%) of sp^3 bonding. *Savvides* and *Bell* [38.94] reported an increase in hardness from 12 to 30 GPa and an increase in elastic modulus from 62 to 213 GPa with an increase in the sp^3-to-sp^2 bonding ratio, from 3 to 6, for a C:H coating deposited by low-energy ion-assisted unbalanced magnetron sputtering of a graphite target in an Ar-H_2 mixture.

Bhushan et al. [38.35] reported hardnesses of about 15 and 35 GPa and elastic moduli of about 140 and 200 GPa, measured by nanoindentation, for a-C:H coatings deposited by DC magnetron sputtering and RF-plasma-enhanced chemical vapor deposition techniques, respectively. The high hardness of RF-PECVD a-C:H coatings is attributed to a higher concentration of sp^3 bonding than in a sputtered hydrogenated a-C:H coating. Hydrogen is believed to play a crucial role in the bonding configuration of carbon atoms by helping

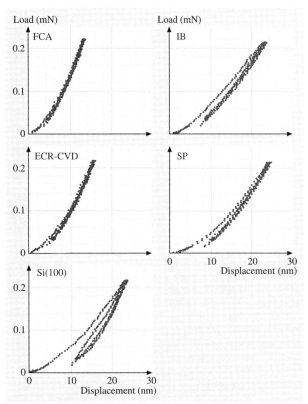

Fig. 38.9 Load versus displacement plots for various 100 nm thick amorphous carbon coatings on single-crystal silicon substrate and bare substrate

Table 38.5 Hardness, elastic modulus, fracture toughness, fatigue life, critical load during scratch, coefficient of friction during accelerated wear testing and residual stress for various DLC coatings on single-crystal silicon substrate

Coating	Hardness[a] [38.48] (GPa)	Elastic modulus[a] [38.48] (GPa)	Fracture toughness[a] [38.89] (MPa m$^{1/2}$)	Fatigue life[b] N_f[d] [38.90] ×10^4	Critical load during scratch[b] [38.48] (mN)	Coefficient of friction during accelerated wear testing[b] [38.48]	Compressive residual stress[c] [38.47] (GPa)
Cathodic arc carbon coating (a-C)	24	280	11.8	2.0	3.8	0.18	12.5
Ion beam carbon coating (a-C:H)	19	140	4.3	0.8	2.3	0.18	1.5
ECR-CVD carbon coating (a-C:H)	22	180	6.4	1.2	5.3	0.22	0.6
DC sputtered carbon coating (a-C:H)	15	140	2.8	0.2	1.1	0.32	2.0
Bulk graphite (for comparison)	Very soft		9–15	–	–	–	–
Diamond (for comparison)	80–104	900–1050	–	–	–	–	–
Si(100) substrate	11	220	0.75	–	0.6	0.55	0.02

[a] Measured on 100 nm thick coatings
[b] Measured on 20 nm thick coatings
[c] Measured on 400 nm thick coatings
[d] N_f was obtained for a mean load of 10 μN and a load amplitude of 8 μN

to stabilize the tetrahedral coordination (sp^3 bonding) of carbon species. *Jansen* et al. [38.78] suggested that the incorporation of hydrogen efficiently passivates the dangling bonds and saturates the graphitic bonding to some extent. However, a large concentration of hydrogen in the plasma in sputter deposition is undesirable. *Cho* et al. [38.33] and *Rubin* et al. [38.40] observed that the hardness decreased from 15 to 3 GPa with increased hydrogen content. *Bhushan* and *Doerner* [38.95] reported a hardness of about 10–20 GPa and an elastic modulus of about 170 GPa, measured by nanoindentation, for 100 nm thick DC magnetron sputtered a-C:H on the silicon substrate.

Residual stresses measured using a well-known curvature measurement technique are also presented in Table 38.5. The DLC coatings are under significant compressive internal stresses. Very high compressive stresses in FCA coatings are believed to be partly responsible for their high hardness. However, high stresses result in coating delamination and buckling. For this reason, the coatings that are thicker than about 1 μm have a tendency to delaminate from the substrates.

Fracture Toughness
Representative load–displacement curves of indentations on 400 nm thick cathodic arc carbon coating on silicon for various peak loads are shown in Fig. 38.11. Steps are found in all of the curves, as shown by arrows in Fig. 38.11a. In the 30 mN SEM micrograph, in addition to several radial cracks, ring-like through-thickness cracking is observed with small lips of material overhanging the edge of indentation. The steps at about 23 mN in the loading curves of indentations made with 30 and 100 mN peak indentation loads result from the ring-like through-thickness cracking. The step at 175 mN in the loading curve of the indentation made

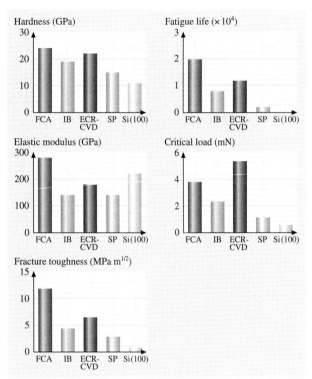

Fig. 38.10 Bar charts summarizing data for various coatings and single-crystal silicon substrate. Hardnesses, elastic moduli, and fracture toughnesses were measured on 100 nm thick coatings, and fatigue lifetimes and critical loads during scratch were measured on 20 nm thick coatings

with 200 mN peak indentation load is caused by spalling and second ring-like through-thickness cracking.

According to *Li* et al. [38.87], the fracture process progresses in three stages: (1) ring-like through-thickness cracks form around the indenter due to high stresses in the contact area, (2) delamination and buckling occur around the contact area at the coating–substrate interface due to high lateral pressure, and (3) second ring-like through-thickness cracks and spalling are generated by high bending stresses at the edges of the buckled coating (Fig. 38.12a). In the first stage, if the coating under the indenter is separated from the bulk coating via the first ring-like through-thickness cracking, a corresponding step will be present in the loading curve. If discontinuous cracks form and the coating under the indenter is not separated from the remaining coating, no step appears in the loading curve, because the coating still supports the indenter and the indenter cannot suddenly advance into the material. In the second stage, for the coating used in the present study, the advances of the indenter during the radial cracking delamination, and buckling are not big enough to form steps in the loading curve, because the coating around the indenter still supports the indenter, but they generate discontinuities that change the slope of the loading curve with increasing indentation load. In the third stage, the stress concentration at the end of the interfacial crack cannot be relaxed by the propagation of the interfacial crack. With an increase in indentation depth, the height of the bulged coating increases. When the height reaches a critical value, the bending stresses caused by the bulged coating around the indenter will result in second ring-like through-thickness crack formation and spalling at the edge of the buckled coating, as shown in Fig. 38.12a, which leads to a step in the loading curve. This is a single event and it results in the separation of the part of the coating around the indenter from the bulk coating via cracking through coatings. The step in the loading curve results (completely) from the coating cracking and not from the interfacial cracking or the substrate cracking.

The area under the load–displacement curve is the work performed by the indenter during the elastic–plastic deformation of the coating/substrate system. The strain energy release in the first/second ring-like cracking and spalling can be calculated from the corresponding steps in the loading curve. Figure 38.12b shows a modeled load–displacement curve. 0ACD is the loading curve and DE is the unloading curve. The first ring-like through-thickness crack should be considered. It should be emphasized that the edge of the buckled coating is far from the indenter and, therefore, it does not matter if the indentation depth exceeds the coating thickness, or if deformation of the substrate occurs around the indenter when we measure the fracture toughness of the coating from the energy released during the second ring-like through-thickness cracking (spalling). Suppose that the second ring-like through-thickness cracking occurs at AC. Now, let us consider the loading curve 0AC. If the second ring-like through-thickness crack does not occur, 0A will extend to 0B to reach the same displacement as 0C. This means that crack formation changes the loading curve 0AB into 0AC. For point B, the elastic–plastic energy stored in the coating/substrate system should be 0BF. For point C, the elastic–plastic energy stored in the coating/substrate system should be 0ACF. Therefore, the energy difference before and after the crack generation is the area of ABC, so this energy stored in ABC will be released as strain energy, creating the ring-like through-thickness crack. According to the theoretical analysis by Li et al. [38.87], the fracture toughness of a thin film can be written as

$$K_{Ic} = \left[\left(\frac{E}{(1-\nu^2)\,2\pi C_R}\right)\left(\frac{U}{t}\right)\right]^{1/2}, \quad (38.2)$$

where E is the elastic modulus, ν is the Poisson ratio, $2\pi C_R$ is the crack length in the coating plane, t is the coating thickness, and U is the strain energy difference before and after cracking.

The fracture toughness of the coatings can be calculated using (38.2). The loading curve is extrapolated along the tangential direction of the loading curve from the starting point of the step up to reach the same displacement as the step. The area between the extrapolated line and the step is the estimated strain energy difference before and after cracking. C_R is measured from SEM micrographs or AFM images of indentations. The second ring-like crack is where the spalling occurs. For example, for the 400 nm thick cathodic arc carbon coating data presented in Fig. 38.11, the U value of 7.1 nN m is assessed from the steps in Fig. 38.11a at peak indentation loads of 200 mN. For a C_R value of 7.0 μm, from Fig. 38.11b, with $E = 300$ GPa (measured using a nanoindenter and an assumed value of 0.25 for ν), fracture toughness values are calculated as 10.9 MPa $\sqrt{m}$ [38.87, 88]. The fracture toughness and related data for various 100 nm thick DLC coatings are presented in Fig. 38.10 and Table 38.5.

Nanofatigue

Delayed fracture resulting from extended service is called fatigue [38.96]. Fatigue fracturing progresses through a material via changes within the material at the tip of a crack, where there is a high stress intensity. There are several situations: cyclic fatigue, stress corrosion and static fatigue. Cyclic fatigue results from cyclic loading of machine components. In a low-flying slider in a magnetic head-disk interface, isolated asperity contacts occur during use and the fatigue failure occurs in the multilayered thin film structure of the magnetic disk [38.13]. Impact occurs in many MEMS components and the failure mode is cyclic fatigue. Asperity contacts can be simulated using a sharp diamond tip in oscillating contact with the component.

Figure 38.13 shows the schematic of a fatigue test on a coating/substrate system using a continuous stiffness measurement (CSM) technique. Load cycles are applied to the coating, resulting in cyclic stress. P is the

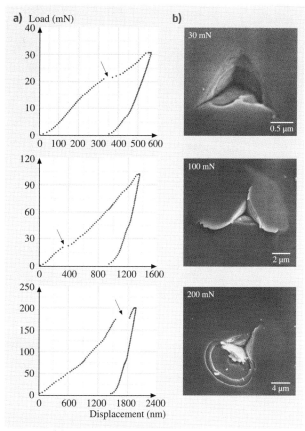

Fig. 38.11 (a) Load–displacement curves of indentations made with 30, 100, and 200 mN peak indentation loads using the cube corner indenter, and (b) SEM micrographs of indentations on a 400 nm thick cathodic arc carbon coating on silicon. Arrows indicate steps during the loading portion of the load–displacement curve [38.87]

cyclic load, P_{mean} is the mean load, P_0 is the oscillation load amplitude, and ω is the oscillation frequency. The following results can be obtained: (1) endurance limit (the maximum load below which there is no coating failure for a preset number of cycles); (2) number of cycles at which the coating failure occurs; and (3) changes in contact stiffness (measured using the unloading slope of each cycle), which can be used to monitor the propagation of interfacial cracks during a cyclic fatigue process.

Figure 38.14a shows the contact stiffness as a function of the number of cycles for 20 nm thick FCA coatings cyclically deformed by various oscillation load amplitudes with a mean load of 10 μN at a frequency of

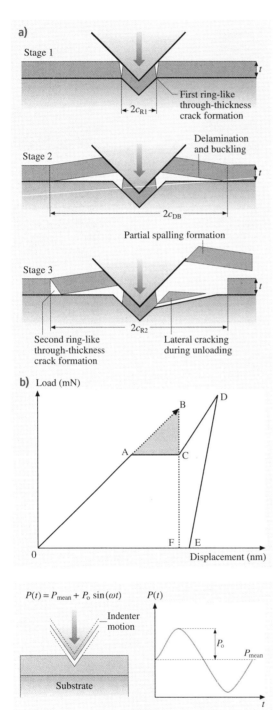

Fig. 38.12 (a) Schematic of various stages in nanoindentation fracture for the coating/substrate system, and (b) schematic of a load–displacement curve showing a step during the loading cycle and the associated energy release ◂

45 Hz. At 4 μN load amplitude, no change in contact stiffness was found for all coatings. This indicates that 4 μN load amplitude is not high enough to damage the coatings. At 6 μN load amplitude, an abrupt decrease in contact stiffness was found after a certain number of cycles for each coating, indicating that fatigue damage had occurred. With increasing load amplitude, the number of cycles to failure N_f decreases for all coatings. Load amplitude versus N_f, a so-called S–N curve, is plotted in Fig. 38.14b. The critical load amplitude below which no fatigue damage occurs (an endurance limit), was identified for each coating. This critical load amplitude, together with the mean load, are of critical importance to the design of head-disk interfaces or MEMS/NEMS device interfaces.

To compare the fatigue lives of the different coatings studied, the contact stiffness is shown as a function of the number of cycles for 20 nm thick FCA, IB, ECR-CVD and SP coatings cyclically deformed by an oscillation load amplitude of 8 μN with a mean load of 10 μN at a frequency of 45 Hz in Fig. 38.14c. The FCA coating has the largest N_f, followed by the ECR-CVD, IB and SP coatings. In addition, after N_f, the contact stiffness of the FCA coating shows a slower decrease than the other coatings. This indicates that the FCA coating was less damaged than the others after N_f. The fatigue behaviors of FCA and ECR-CVD coatings of different thicknesses are compared in Fig. 38.14d. For both coatings, N_f decreases with decreasing coating thickness. At 10 nm, FCA and ECR-CVD have almost the same fatigue life. At 5 nm, the ECR-CVD coating shows a slightly longer fatigue life than the FCA coating. This indicates that the microstructure and residual stresses are not uniform across the thickness direction, even for nanometer-thick DLC coatings. Thinner coatings are more influenced by interfacial stresses than thicker coatings.

Figure 38.15a shows high-magnification SEM images of 20 nm thick FCA coatings before, at, and after N_f. In the SEM images, the net-like structure is the gold film coated on the DLC coating, which should be

Fig. 38.13 Schematic of a fatigue test on a coating/substrate system using the continuous stiffness measurement technique ◂

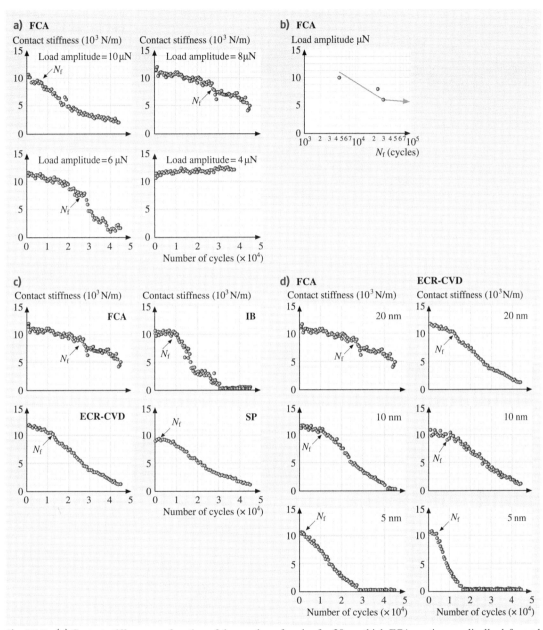

Fig. 38.14 (a) Contact stiffness as a function of the number of cycles for 20 nm thick FCA coatings cyclically deformed by various oscillation load amplitudes with a mean load of 10 μN at a frequency of 45 Hz; (b) plot of load amplitude versus N_f; (c) contact stiffness as a function of the number of cycles for four different 20 nm thick coatings with a mean load of 10 μN and a load amplitude of 8 μN at a frequency of 45 Hz; and (d) contact stiffness as a function of the number of cycles for two coatings of different thicknesses at a mean load of 10 μN and a load amplitude of 8 μN at a frequency of 45 Hz

ignored when analyzing the indentation fatigue damage. Before N_f, no delamination or buckling was found except for the residual indentation mark at magnifications of up to $1\,200\,000\times$ using SEM. This suggests that only plastic deformation occurred before N_f. At N_f, the coating around the indenter bulged upwards, indicating delamination and buckling. Therefore, it is believed that the decrease in contact stiffness at N_f results from delamination and buckling of the coating from the substrate. After N_f, the buckled coating was broken down around the edge of the buckled area, forming a ring-like crack. The remaining coating overhung at the edge of the buckled area. It is noted that the indentation size increases with the number of cycles. This indicates that deformation, delamination and buckling as well as ring-like crack formation occurred over a period of time.

The schematic in Fig. 38.15b shows various stages of indentation fatigue damage for a coating/substrate system. Based on this study, three stages of indentation fatigue damage appear to exist: (1) indentation-induced compression; (2) delamination and buckling; (3) ring-like crack formation at the edge of the buckled coating.

The deposition process often induces residual stresses in coatings. The model shown in Fig. 38.15b considers a coating with uniform biaxial residual compression σ_r. In the first stage, indentation induces elastic/plastic deformation, exerting a pressure (acting outward) on the coating around the indenter. Interfacial defects like voids and impurities act as original cracks. These cracks propagate and link up as the indentation compressive stress increases. At this stage, the coating, which is under the indenter and above the interfacial crack (with a crack length of $2a$), still maintains a solid contact with the substrate; the substrate still fully supports the coating. Therefore, this interfacial crack does not lead to an abrupt decrease in contact stiffness, but gives rise to a slight decrease in contact stiffness, as shown in Fig. 38.14. The coating above the interfacial crack is treated as a rigidly clamped disk. We assume that the crack radius a is large compared with the coating thickness t. Since the coating thickness ranges from 5 to 20 nm, this assumption is easily satisfied in this study (the radius of the delaminated and buckled area, shown in Fig. 38.15a, is on the order of 100 nm). The compressive stress caused by indentation is given as [38.97]

$$\sigma_i = \frac{E}{(1-\nu)}\varepsilon_i = \frac{EV_i}{2\pi t a^2(1-\nu)}, \qquad (38.3)$$

where ν and E are the Poisson ratio and elastic modulus of the coating, V_i is the indentation volume, t is the coating thickness, and a is the crack radius. As the number of cycles increases, so does the indentation volume V_i. Therefore, the indentation compressive stress σ_i increases accordingly. In the second stage, buckling occurs during the unloading segment of the fatigue testing cycle when the sum of the indentation compressive stress σ_i and the residual stress σ_r exceed the critical buckling stress σ_b for the delaminated circular section, as given by [38.98]

$$\sigma_b = \frac{\mu^2 E}{12(1-\nu^2)}\left(\frac{t}{a}\right)^2, \qquad (38.4)$$

where the constant μ equals 42.67 for a circular clamped plate with a constrained center point and 14.68 when the center is unconstrained. The buckled coating acts as a cantilever. In this case, the indenter indents a cantilever rather than a coating/substrate system. This ultrathin coating cantilever has much less contact stiffness than the coating/substrate system. Therefore, the contact stiffness shows an abrupt decrease at N_f. In

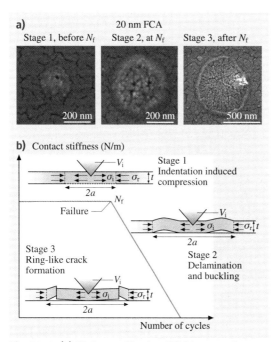

Fig. 38.15 (a) High-magnification SEM images of a coating before, at, and after N_f, and **(b)** schematic of various stages of indentation fatigue damage for a coating/substrate system [38.90]

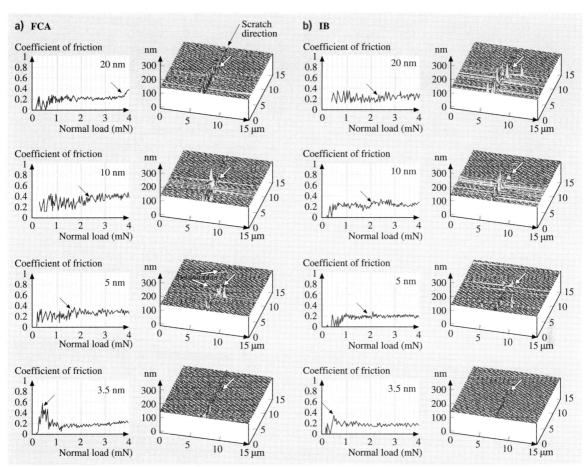

Fig. 38.16a–d Coefficient of friction profiles as a function of normal load, as well as corresponding AFM surface height maps of regions over scratches at the respective critical loads (indicated by the *arrows* in the friction profiles and AFM images), for coatings of different thicknesses deposited by various deposition techniques: (**a**) FCA, (**b**) IB

the third stage, with more cycles, the delaminated and buckled size increases, resulting in a further decrease in contact stiffness since the cantilever beam length increases. On the other hand, a high bending stress acts at the edge of the buckled coating. The larger the buckled size, the higher the bending stress. The cyclic bending stress causes fatigue damage at the end of the buckled coating, forming a ring-like crack. The coating under the indenter is separated from the bulk coating (caused by the ring-like crack at the edge of the buckled coating) and the substrate (caused by the delamination and buckling in the second stage). Therefore, the coating under the indenter is not constrained; it is free to move with the indenter during fatigue test-

ing. At this point, the sharp nature of the indenter is lost, because the coating under the indenter gets stuck on the indenter. The indentation fatigue experiment results in the contact of a (relatively) huge blunt tip with the substrate. This results in a low contact stiffness value.

Compressive residual stresses result in delamination and buckling. A coating with a higher adhesion strength and a lower compressive residual stress is required for a higher fatigue life. Interfacial defects should be avoided in the coating deposition process. We know that ring-like crack formation occurs in the coating. Formation of fatigue cracks in the coating depends upon the hardness and the fracture toughness. Cracks are

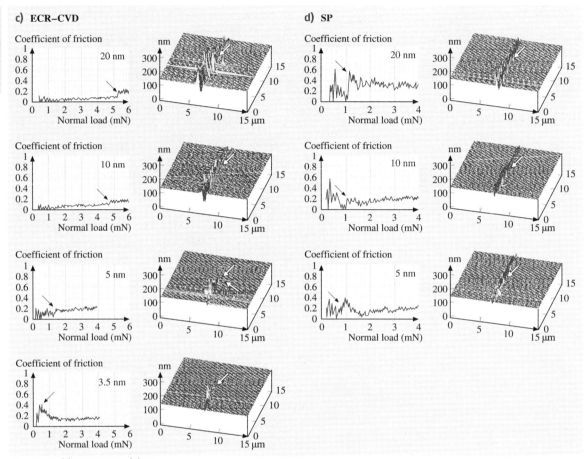

Fig. 38.16 (c) ECR-CVD, (d) SP

more difficult to form and propagate in the coating with higher strength and fracture toughness.

It is now accepted that long fatigue life in a coating/substrate almost always involves *living with a crack*, that the threshold or limit condition is associated with the nonpropagation of existing cracks or defects, even though these cracks may be undetectable [38.96]. For all of the coatings studied at 4 μN, the contact stiffness does not change much. This indicates that delamination and buckling did not occur within the number of cycles tested in this study. This is probably because the indentation-induced compressive stress was not high enough to allow the cracks to propagate and link up under the indenter, or the sum of the indentation compressive stress σ_i and the residual stress σ_r did not exceed the critical buckling stress σ_b.

Figure 38.10 and Table 38.5 summarize the hardnesses, elastic moduli, fracture toughnesses, and fatigue lifetimes of all of the coatings studied. A good correlation exists between the fatigue life and other mechanical properties. Higher mechanical properties result in a longer fatigue life. The mechanical properties of DLC coatings are controlled by the sp^3-to-sp^2 ratio. An sp^3-bonded carbon exhibits the outstanding properties of diamond [38.51]. Higher kinetic energy during deposition will result in a larger fraction of sp^3-bonded carbon in an amorphous network. Thus, the higher kinetic energy for the FCA could be responsible for its enhanced carbon structure and mechanical properties [38.48–50,99]. Higher adhesion strength between the FCA coating and substrate makes the FCA coating more difficult to delaminate from the substrate.

38.3.2 Microscratch and Microwear Studies

For microscratch studies, a conical diamond indenter (that has a tip radius of about 1 μm and a cone angle of 60° for example) is drawn over the sample surface, and the load is ramped up (typically from 2 to 25 mN) until substantial damage occurs. The coefficient of friction is monitored during scratching. Scratch-induced coating damage, specifically fracture or delamination, can be monitored by in situ friction force measurements and using optical and SEM imaging of the scratches after tests. A gradual increase in friction is associated with plowing, and an abrupt increase in friction is associated with fracture or catastrophic failure [38.100]. The load corresponding to an abrupt increase in friction or an increase in friction above a certain value (typically 2× the initial value) provides a measure of the scratch resistance or the adhesive strength of a coating, and is called the *critical load*. The depths of scratches are measured with increasing scratch length or normal load using an AFM, typically with an area of $10 \times 10\,\mu\text{m}^2$ [38.48, 49, 101].

Microscratch and microwear studies are also conducted using an AFM [38.23, 50, 99, 102, 103]. A square pyramidal diamond tip (tip radius ≈ 100 nm) or a three-sided pyramidal diamond (Berkovich) tip with an apex angle of 60° and a tip radius of about 100 nm, mounted on a platinum-coated, rectangular stainless steel cantilever of stiffness of about 40 N/m, is scanned orthogonal to the long axis of the cantilever to generate scratch and wear marks. During the scratch test, the normal load is either kept constant or is increased (typically from 0 to 100 μN) until damage occurs. Topographical images of the scratch are obtained in situ with the AFM at a low load. By scanning the sample during scratching, wear experiments can be conducted. Wear is monitored as a function of the number of cycles at a constant load. Normal loads (10–80 μN) are typically used.

Microscratch

Scratch tests conducted with a sharp diamond tip simulate a sharp asperity contact. In a scratch test, the cracking or delamination of a hard coating is signaled by a sudden increase in the coefficient of friction [38.23]. The load associated with this event is called the *critical load*.

Wu [38.104], *Bhushan* et al. [38.70], *Gupta* and *Bhushan* [38.12, 47], and *Li* and *Bhushan* [38.48, 49, 101] have used a nanoindenter to perform microscratch (mechanical durability) studies of various carbon coatings. The coefficient of friction profiles as a function

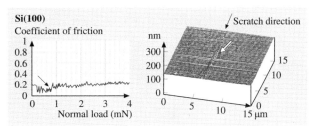

Fig. 38.17 Coefficient of friction profiles as a function of normal load as well as corresponding AFM surface height maps of regions over scratches at the respective critical loads (indicated by the *arrows* in the friction profiles and AFM images) for Si(100)

of increasing normal load as well as AFM surface height maps of regions over scratches at the respective critical loads (indicated by the arrows in the friction profiles and AFM images) observed for coatings with different thicknesses and single-crystal silicon substrate using a conical tip are compared in Figs. 38.16 and 38.17. *Bhushan* and *Koinkar* [38.102], *Koinkar* and *Bhushan* [38.103], *Bhushan* [38.23], and *Sundararajan* and *Bhushan* [38.50, 99] used an AFM to perform microscratch studies. Data obtained for coatings with different thicknesses and silicon substrate using a Berkovich tip are compared in Figs. 38.18 and 38.19. Critical loads for various coatings tested using a nanoindenter and AFM are summarized in Fig. 38.20. Selected data for 20 nm thick coatings obtained using nanoindenter are also presented in Fig. 38.10 and Table 38.5.

It is clear that a well-defined critical load exists for each coating. The AFM images clearly show that below the critical loads the coatings were plowed by the scratch tip, associated with the plastic flow of materials. At and after the critical loads, debris (chips) or buckling was observed on the sides of the scratches. Delamination or buckling can be observed around or after the critical loads, which suggests that the damage starts from delamination and buckling. For the 3.5 and 5 nm thick FCA coatings, before the critical loads small debris is observed on the sides of the scratches. This suggests that the thinner FCA coatings are not so durable. It is obvious that, for a given deposition method, the critical loads increase with increasing coating thickness. This indicates that the critical load is determined not only by the strength of adhesion to the substrate, but also by the coating thickness. We note that more debris generated on the thicker coatings than thinner coatings. A thicker coating is more difficult to break; the broken coating chips (debris) seen for a thicker coat-

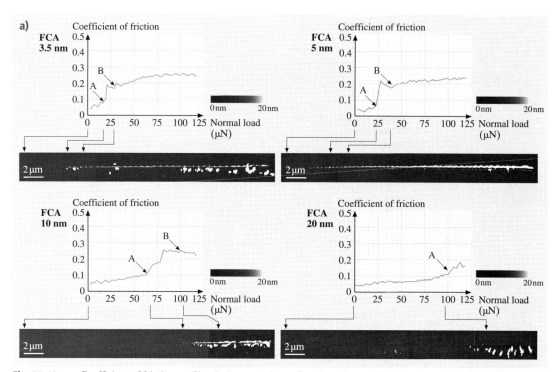

Fig. 38.18a–c Coefficient of friction profiles during scratch as a function of normal load and corresponding AFM surface height maps for (**a**) FCA, (**b**) ECR-CVD, and (**c**) SP coatings [38.99] ▲ ▶

ing are larger than those for the thinner coatings. The different residual stresses of coatings of different thicknesses may also affect the size of the debris. The AFM image shows that the silicon substrate was damaged by plowing, associated with the plastic flow of material. At

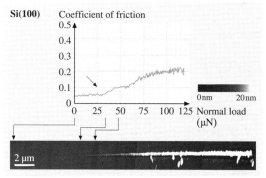

Fig. 38.19 Coefficient of friction profiles during scratch as a function of normal load and corresponding AFM surface height maps for Si(100) [38.99]

and after the critical load, a small amount of uniform debris is observed and the amount of debris increases with increasing normal load.

Since the damage mechanism at the critical load appears to be the onset of plowing, harder coatings with more fracture toughness will therefore require a higher load for deformation and hence a higher critical load. Figure 38.21 gives critical loads of various coatings, obtained with AFM tests, as a function of the coating hardness and fracture toughness (from Table 38.5). It can be seen that, in general, increasing coating hardness and fracture toughness results in a higher critical load. The only exceptions are the FCA coatings at 5 and 3.5 nm thickness, which show the lowest critical loads despite their high hardness and fracture toughness. The brittleness of the thinner FCA coatings may be one reason for their low critical loads. The mechanical properties of coatings that are less than 10 nm thick are not known. The FCA process may result in coatings with low hardness at low thickness due to differences in coating stoichiometry and structure compared to coatings of higher thickness. Also, at these thicknesses stresses at

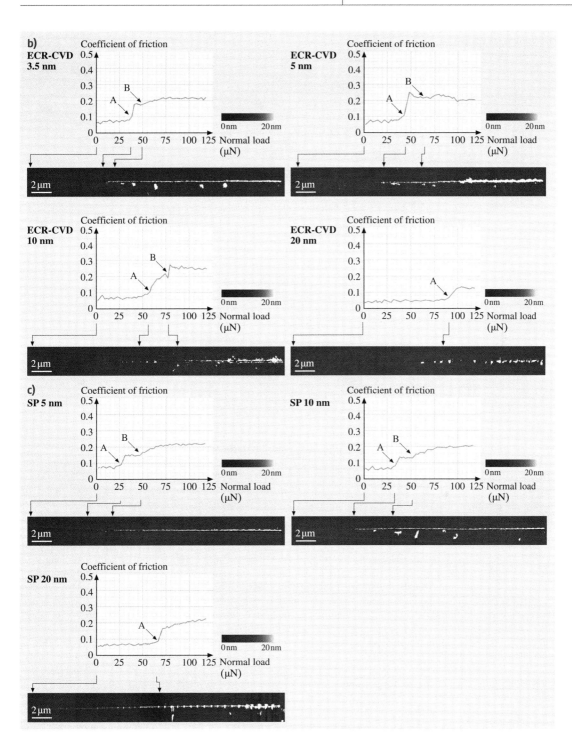

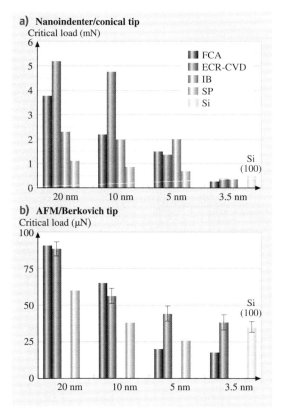
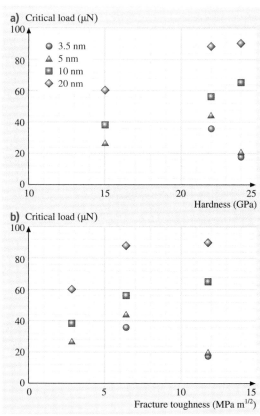

Fig. 38.20a,b Critical loads estimated from the coefficient of friction profiles from (**a**) nanoindenter and (**b**) AFM tests for various coatings of different thicknesses and Si(100) substrate

Fig. 38.21a,b Measured critical loads estimated from the coefficient of friction profiles from AFM tests as a function of (**a**) coating hardness and (**b**) fracture toughness. Coating hardness and fracture toughness values were obtained using a nanoindenter on 100 nm thick coatings (Table 38.5)

the coating–substrate interface may affect coating adhesion and load-carrying capacity.

Based on the experimental results, a schematic of the scratch damage mechanisms encountered for the DLC coatings used in this study is shown in Fig. 38.22. Below the critical load, if a coating has a good combination of strength and fracture toughness, plowing associated with the plastic flow of materials is responsible for the coating damage (Fig. 38.22a). However, if the coating has a low fracture toughness, cracking could occur during plowing, resulting in the formation of small amounts of debris (Fig. 38.22b). When the normal load is increased to the critical load, delamination or buckling will occur at the coating–substrate interface (Fig. 38.22c). A further increase in normal load will result in coating breakdown via through-coating thickness cracking, as shown in Fig. 38.22d. Therefore,

adhesion strength plays a crucial role in the determination of critical load. If a coating adheres strongly to the substrate, the coating is more difficult to delaminate, which will result in a higher critical load. The interfacial and residual stresses of a coating may also greatly affect the delamination and buckling [38.1]. A coating with higher interfacial and residual stresses is more easily delaminated and buckled, which will result in a low critical load. It was reported earlier that FCA coatings have higher residual stresses than other coatings [38.47]. Interfacial stresses play an increasingly important role as the coating gets thinner. A large mismatch in elastic modulus between the FCA coating and the silicon substrate may cause large interfacial stresses. This may be why thinner FCA coatings show lower crit-

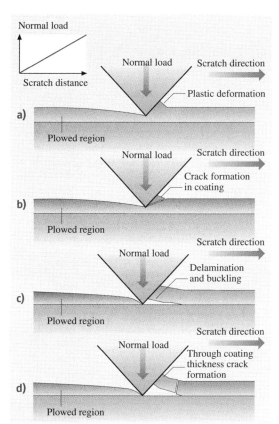

Fig. 38.22a–d Schematic of scratch damage mechanisms for DLC coatings: (**a**) plowing associated with the plastic flow of materials, (**b**) plowing associated with the formation of small debris, (**c**) delamination and buckling at the critical load, and (**d**) breakdown via through-coating thickness cracking at and after the critical load [38.48]

ical loads than thicker FCA coatings, even though the FCA coatings have higher hardness and elastic moduli. The brittleness of thinner FCA coatings may be another reason for the lower critical loads. The strength and fracture toughness of a coating also affect the critical load. Greater strength and fracture toughness will make the coating more difficult to break after delamination and buckling. The high scratch resistance/adhesion of FCA coatings is attributed to the atomic intermixing that occurs at the coating–substrate interface due to the high kinetic energy (2 keV) of the plasma formed during the cathodic arc deposition process [38.57]. This atomic intermixing provides a graded compositional transition between the coating and the substrate material. In all other coatings used in this study, the kinetic energy of the plasma is insufficient for atomic intermixing.

Gupta and *Bhushan* [38.12, 47] and *Li* and *Bhushan* [38.48, 49] measured the scratch resistances of DLC coatings deposited on Al_2O_3-TiC, Ni-Zn ferrite and single-crystal silicon substrates. An interlayer of silicon is required to adhere the DLC coating to other substrates, except in the case of cathodic arc-deposited coatings. The best adhesion with cathodic arc carbon coating is obtained on electrically conducting substrates such as Al_2O_3-TiC and silicon rather than Ni-Zn ferrite.

Microwear

Microwear studies can be conducted using an AFM [38.23]. For microwear studies, a three-sided pyramidal single-crystal natural diamond tip with an apex angle of about $80°$ and a tip radius of about 100 nm is used at relatively high loads of $1-150\,\mu N$. The diamond tip is mounted on a stainless steel cantilever beam with a normal stiffness of about 30 N/m. The sample is generally scanned in a direction orthogonal to the long axis of the cantilever beam (typically at a rate of 0.5 Hz). The tip is mounted on the beam such that one of its edges is orthogonal to the beam axis. In wear studies, an area of $2 \times 2\,\mu m^2$ is typically scanned for a selected number of cycles.

Microwear studies of various types of DLC coatings have been conducted [38.50, 102, 103]. Figure 38.23 a shows a wear mark on uncoated Si(100). Wear occurs uniformly and material is removed layer-by-layer via plowing from the first cycle, resulting in constant friction force during the wear (Fig. 38.24a). Figure 38.23b shows AFM images of the wear marks on 10 nm thick coatings. It is clear that the coatings wear nonuniformly. Coating failure is sudden and accompanied by a sudden rise in the friction force (Fig. 38.24b). Figure 38.24 shows the wear depth of Si(100) substrate and various DLC coatings at two different loads. FCA- and ECR-CVD-deposited 20 nm thick coatings show excellent wear resistance up to $80\,\mu N$, the load that is required for the IB 20 nm coating to fail. In these tests, *failure* of a coating occurs when the wear depth exceeds the quoted coating thickness. The SP 20 nm coating fails at a much lower load of $35\,\mu N$. At $60\,\mu N$, the coating hardly provides any protection. Moving on to the 10 nm coatings, the ECR-CVD coating requires about 45 cycles at $60\,\mu N$ to fail, whereas the IB and FCA, coatings fail at $45\,\mu N$. The FCA coating exhibits slight roughening in the wear track after the first few cycles, which leads to an increase in the friction force. The SP coating continues to exhibit poor resistance, failing

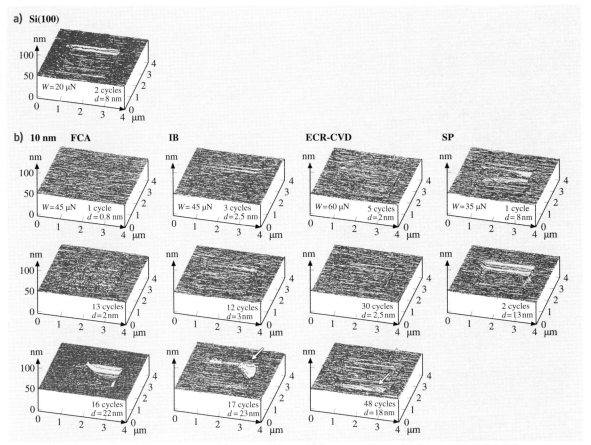

Fig. 38.23a,b AFM images of wear marks on (**a**) bare Si(100), and (**b**) various 10 nm thick DLC coatings [38.50]

at 20 μN. For the 5 nm coatings, the load required to make the coatings fail continues to decrease, but IB and ECR-CVD still provide adequate protection compared to bare Si(100) in that order, with the silicon failing at 35 μN, the FCA coating at 25 μN and the SP coating at 20 μN. Almost all of the 20, 10, and 5 nm coatings provide better wear resistance than bare silicon. At 3.5 nm, FCA coating provides no wear resistance, failing almost instantly at 20 μN. The IB and ECR-CVD coatings show good wear resistance at 20 μN compared to bare Si(100). However, IB only lasts about ten cycles and ECR-CVD about three cycles at 25 μN.

The wear tests highlight the differences between the coatings more vividly than the scratch tests. At higher thicknesses (10 and 20 nm), the ECR-CVD and FCA coatings appear to show the best wear resistance. This is probably due to the greater hardness of the coatings (Table 38.5). At 5 nm, the IB coating appears to be the best. FCA coatings show poorer wear resistance with decreasing coating thickness. This suggests that the trends in hardness seen in Table 38.5 no longer hold at low thicknesses. SP coatings show consistently poor wear resistance at all thicknesses. The 3.5 nm IB coating does provide reasonable wear protection at low loads.

38.3.3 Macroscale Tribological Characterization

So far, we have presented data on mechanical characterization and microscratch and microwear studies using a nanoindenter and an AFM. Mechanical properties affect the tribological performance of a coating, and microwear studies simulate a single asperity contact, which helps us to understand the wear process.

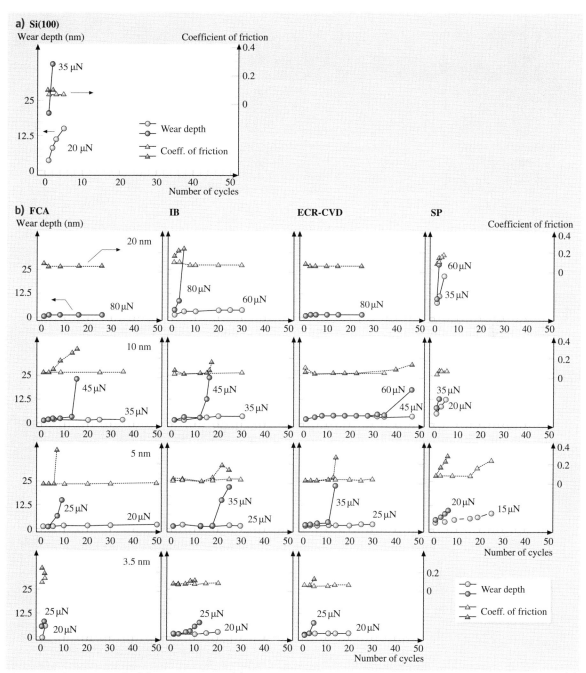

Fig. 38.24a,b Wear data for (**a**) bare Si(100) and (**b**) various DLC coatings. Coating thickness is constant along each row in (**b**). Both the wear depth and the coefficient of friction during wear are plotted for a given cycle [38.50]

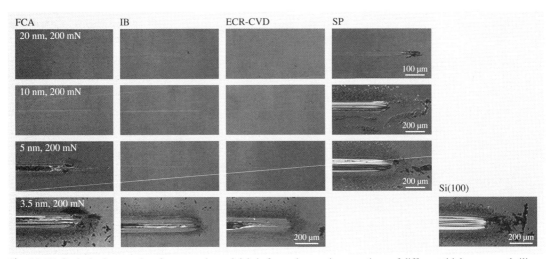

Fig. 38.25 Optical micrographs of wear tracks and debris formed on various coatings of different thicknesses and silicon substrate when slid against a sapphire ball after a sliding distance of 5 nm. The end of the wear track is on the right-hand side of the image

These studies are useful when screening various candidate coatings, and also aid our understanding of the relationships between deposition conditions and properties of the samples. In the next step, macroscale friction and wear tests need to be conducted to measure the tribological performance of the coating.

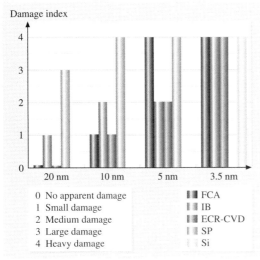

Fig. 38.26 Bar chart of the wear damage indices for various coatings of different thicknesses and Si(100) substrate based on optical examination of the wear tracks and debris

Macroscale accelerated friction and wear tests have been conducted to screen a large number of candidates, as have functional tests on selected candidates. An accelerated test is designed to accelerate the wear process such that it does not change the failure mechanism. The accelerated friction and wear tests are generally conducted using a ball-on-flat tribometer under reciprocating motion [38.70]. Typically, a diamond tip with a tip radius of 20 μm or a sapphire ball with a diameter of 3 mm and a surface finish of about 2 nm RMS is slid against the coated substrates at selected loads. The coefficient of friction is monitored during the tests.

Functional tests are conducted using an actual machine running at close to the actual operating conditions for which the coatings have been developed. The tests are generally accelerated somewhat to fail the interface in a short time.

Accelerated Friction and Wear Tests

Li and *Bhushan* [38.48] conducted accelerated friction and wear tests on DLC coatings deposited by various deposition techniques using a ball-on-flat tribometer. The average coefficient of friction values observed are presented in Table 38.5. Optical micrographs of wear tracks and debris formed on all samples when slid against a sapphire ball after a sliding distance of 5 nm are presented in Fig. 38.25. The normal load used for the 20 and 10 nm thick coatings was 200 mN, and the normal load used for the 5 and 3.5 nm thick coatings and the silicon substrate was 150 mN.

Among the 20 nm thick coatings, the SP coating exhibits a higher coefficient of friction (about 0.3) than for the other coatings coefficient of friction (all of which were about 0.2). The optical micrographs show that the SP coating has a larger wear track and more debris than the IB coating. No wear track or debris were found on the 20 nm thick FCA and ECR-CVD coatings. The optical micrographs of 10 nm thick coatings show that the SP coating was severely damaged, showing a large wear track with scratches and lots of debris. The FCA and ECR-CVD coatings show smaller wear tracks and less debris than the IB coatings.

For the 5 nm thick coatings, the wear tracks and debris of the IB and ECR-CVD coatings are comparable. The bad wear resistance of the 5 nm thick FCA coating is in good agreement with the low critical scratch load, which may be due to the higher interfacial and residual stresses as well as the brittleness of the coating.

At 3.5 nm, all of the coatings exhibit wear. The FCA coating provides no wear resistance, failing instantly like the silicon substrate. Large block-like debris is observed on the sides of the wear track of the FCA coating. This indicates that large region delamination and buckling occurs during sliding, resulting in large block-like debris. This block-like debris, in turn, scratches the coating, making the damage to the coating even more severe. The IB and ECR-CVD coatings are able to provide some protection against wear at 3.5 nm.

In order to better evaluate the wear resistance of various coatings, based on an optical examination of the wear tracks and debris after tests, a bar chart of the wear damage index for various coatings of different thicknesses and an uncoated silicon substrate is presented in Fig. 38.26. Among the 20 and 10 nm thick coatings, the SP coatings show the worst damage, followed by FCA/ECR-CVD. At 5 nm, the FCA and SP coatings show the worst damage, followed by the IB and ECR-CVD coatings. All of the 3.5 nm thick coatings show the same heavy damage as the uncoated silicon substrate.

The wear damage mechanisms of the thick and thin DLC coatings studied are believed to be as illustrated in Fig. 38.27. In the early stages of sliding, deformation zone, Hertzian and wear fatigue cracks that have formed beneath the surface extend within the coating upon subsequent sliding [38.1]. Formation of fatigue cracks depends on the hardness and subsequent cycles. These are controlled by the sp^3-to-sp^2 ratio. For thicker coatings, the cracks generally do not penetrate

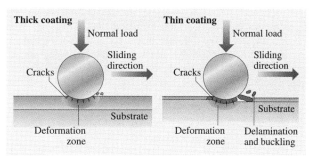

Fig. 38.27 Schematic of wear damage mechanisms for thick and thin DLC coatings [38.48]

the coating. For a thinner coating, the cracks easily propagate down to the interface aided by the interfacial stresses and get diverted along the interface just enough to cause local delamination of the coating. When this happens, the coating experiences excessive plowing. At this point, the coating fails catastrophically, resulting in a sudden rise in the coefficient of friction. All 3.5 nm thick coatings failed much quicker than the thicker coatings. It appears that these thin coatings have very

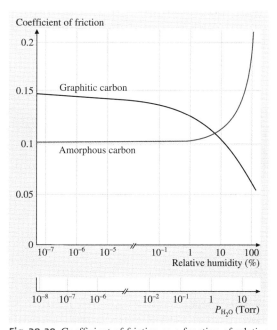

Fig. 38.28 Coefficient of friction as a function of relative humidity and water vapor partial pressure for a RF-plasma deposited amorphous carbon coating and a bulk graphitic carbon coating sliding against a steel ball

low load-carrying capacities and so the substrate undergoes deformation almost immediately. This generates stresses at the interface that weaken the coating adhesion and lead to delamination of the coating. Another reason may be that the thickness is insufficient to produce a coating that has the DLC structure. Instead, the bulk may be made up of a matrix characteristic of the interface region where atomic mixing occurs with the substrate and/or any interlayer used. This would also result in poor wear resistance and silicon-like behavior of the coating, especially for FCA coatings, which show the worst performance at 3.5 nm. SP coatings show the worst wear performance at any thickness (Fig. 38.25). This may be due to their poor mechanical properties, such as lower hardness and scratch resistance, compared to the other coatings.

Comparison of Figs. 38.20 and 38.26 shows a very good correlation between the wear damage and critical scratch loads. Less wear damage corresponds to a higher critical scratch load. Based on the data, thicker coatings do show better scratch and wear resistance than thinner coatings. This is probably due to the better load-carrying capacities of the thick coatings compared to the thinner ones. For a given coating thickness, increased hardness and fracture toughness and better adhesion strength are believed to be responsible for the superior wear performance.

Effect of Environment

The friction and wear performance of an amorphous carbon coating is known to be strongly dependent on the water vapor content and partial gas pressure in the test environment. The friction data for an amorphous carbon film on a silicon substrate sliding against steel are presented as a function of the partial pressure of water vapor in Fig. 38.28 [38.1, 13, 69, 105, 106]. Friction increases dramatically above a relative humidity of about 40%. At high relative humidity, condensed water vapor forms meniscus bridges at the contacting asperities, and the menisci result in an intrinsic attractive force that is responsible for an increase in the friction. For completeness, data on the coefficient of friction of bulk graphitic carbon are also presented in Fig. 38.28. Note that the friction decreases with increased relative humidity [38.107]. Graphitic carbon has a layered crystal lattice structure. Graphite absorbs polar gases (such as H_2O, O_2, CO_2, NH_3) at the edges of the crystallites, which weakens the interlayer bonding forces facilitating interlayer slip and results in lower friction [38.1].

A number of tests have been conducted in controlled environments in order to better study the effects of environmental factors on carbon-coated magnetic disks. *Marchon* et al. [38.108] conducted tests in alternating environments of oxygen and nitrogen gases (Fig. 38.29). The coefficient of friction increases as soon as oxygen is added to the test environment, whereas in a nitrogen environment the coefficient of friction reduces slightly. Tribochemical oxidation of the DLC coating in the oxidizing environment is responsible for an increase in the coefficient of friction, implying wear.

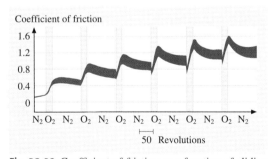

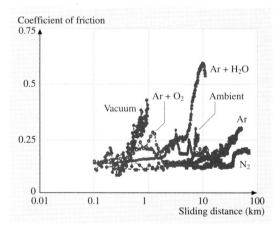

Fig. 38.29 Coefficient of friction as a function of sliding distance for a ceramic slider against a magnetic disk coated with a 20 nm thick DC magnetron sputtered DLC coating, measured at a speed of 0.06 m/s for a load of 10 g. The environment was alternated between oxygen and nitrogen gases [38.108]

Fig. 38.30 Durability, measured by sliding a Al_2O_3-TiC magnetic slider against a magnetic disk coated with a 20 nm thick DC sputtered amorphous carbon coating and 2 nm thick perfluoropolyether film, measured at a speed of 0.75 m/s and for a load of 10 g. Vacuum refers to 2×10^{-7} Torr [38.71]

Dugger et al. [38.109], Strom et al. [38.110], Bhushan and Ruan [38.111] and Bhushan et al. [38.71] conducted tests with DLC-coated magnetic disks (with about 2 nm thick perfluoropolyether lubricant film) in contact with Al_2O_3-TiC sliders in different gaseous environments, including a high vacuum of 2×10^{-7} Torr (Fig. 38.30). The wear lives are the shortest in high vacuum and the longest in atmospheres of mostly nitrogen and argon with the following order (from best to worst): argon or nitrogen, $Ar+H_2O$, ambient, $Ar+O_2$, $Ar+H_2O$, vacuum. From this sequence of wear performance, we can see that having oxygen and water in an operating environment worsens the wear performance of the coating, but having a vacuum is even worse. Indeed, failure mechanisms differ in different environments. In high vacuum, intimate contact between the disk and the slider surface results in significant wear. In ambient air, $Ar+O_2$ and $Ar+H_2O$, tribochemical oxidation of the carbon overcoat is responsible for interface failure. For experiments performed in pure argon and nitrogen, mechanical shearing of the asperities causes the formation of debris, which is responsible for the formation of scratch marks on the carbon surface, which were observed with an optical microscope [38.71].

Functional Tests

Magnetic thin film heads made with Al_2O_3-TiC substrate are used in magnetic storage applications [38.13]. A multilayered thin film pole-tip structure present on the head surface wears more rapidly than the much harder Al_2O_3-TiC substrate. Pole-tip recession (PTR) is a serious concern in magnetic storage [38.15–19, 112]. Two of the diamond-like carbon coatings with superior mechanical properties – ion beam and cathodic arc carbon coatings – were deposited on the air-bearing surfaces of Al_2O_3-TiC head sliders [38.15]. Functional tests were conducted by running a metal particle (MP) tape in a computer tape drive. The average PTR as a function of sliding distance is presented in Fig. 38.31. We note that the PTR increases for the uncoated head, whereas there is a slight increase in PTR for the coated heads during early sliding followed by little change. Thus, the coatings provide protection.

The micromechanical as well as the accelerated and functional tribological data presented here clearly suggest that there is a good correlation between the scratch resistance and wear resistance measured using accelerated tests and functional tests. Thus, scratch tests can be successfully used to screen coatings for wear applications.

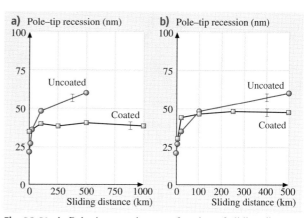

Fig. 38.31a,b Pole-tip recession as a function of sliding distance, measured with an AFM, for (a) uncoated and 20 nm thick ion beam carbon coated, and (b) uncoated and 20 nm thick cathodic arc carbon coated Al_2O_3-TiC heads run against MP tapes [38.15]

38.3.4 Coating Continuity Analysis

Ultrathin (less than 10 nm) coatings may not uniformly coat the sample surface. In other words, the coating may be discontinuous and deposited in the form of islands on the microscale. Therefore, one possible reason for poor wear protection and the nonuniform failure of thin coatings may be poor coverage of the substrate. Coating continuity can be studied using surface analytical techniques such as Auger and/or XPS analyses. Any discontinuity in coating thickness that is less than

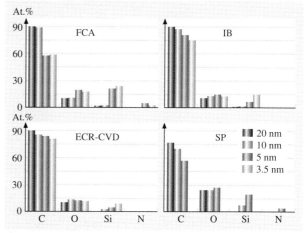

Fig. 38.32 Quantified XPS data for various DLC coatings on Si(100) substrate [38.50]. Atomic concentrations are shown

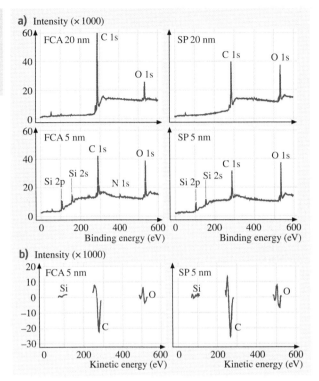

Fig. 38.33 (a) XPS spectra for 5 nm and 20 nm thick FCA and SP coatings on Si(100) substrate, and (b) AES spectra for FCA and SP coatings of 5 nm thickness on Si(100) substrate [38.50]

strates are shown in Fig. 38.32. The sampling depth is about 2–3 nm. The poor SP coatings and the poor 5 nm and 3.5 nm FCA coatings () show much lower carbon contents (atomic concentrations of $< 75\%$ and $< 60\%$ respectively) than the IB and ECR-CVD coatings. Silicon is detected in all of the 5 nm coatings. From the data it is hard to infer whether the Si is from the substrate or from exposed regions due to discontinuous coating. Based on the sampling depth, any Si detected in 3.5 nm coatings would likely be from the substrate. The other interesting observation is that all poor coatings (all SP and FCA 5 and 3.5 nm coatings) have almost twice the oxygen content of the other coatings. Any oxygen present may be due to leaks in the deposition chamber, and it is present as silicon oxides.

AES measurements averaged over a scan area of 900 μm^2 were conducted on FCA and SP 5 nm coatings at six different regions on each sample. Very little silicon was detected on this scale, and the detected peaks were characteristic of oxides. The oxygen levels were comparable to those seen for good coatings via XPS. These results contrast with the XPS measurements performed at a larger scale, suggesting that the coatings only possess discontinuities at isolated areas and that the 5 nm coatings are generally continuous on the microscale. Figure 38.33 shows representative XPS and AES spectra of selected samples.

the sampling depth of the instrument result in the local detection of the substrate species [38.49, 50, 102].

The results from an XPS analysis of 1.3 mm^2 regions (single point measurement with spot diameter of 1300 μm) on various coatings deposited on Si(100) sub-

38.4 Closure

Diamond material and its smooth coatings are used for very low wear and relatively low friction. Major limitations of the true diamond coatings are that they need to be deposited at high temperatures, can only be deposited on selected substrates, and require surface finishing. Hard amorphous carbon (a-C) or commonly known as DLC coatings exhibit mechanical, thermal and optical properties close to that of diamond. These can be deposited with a large range of thicknesses by using a variety of deposition processes, on variety of substrates at or near room temperature. The coatings reproduce substrate topography avoiding the need of post finishing. Friction and wear properties of some DLC coatings can be very attractive for tribological applications. The largest industrial application of these coatings is in magnetic storage devices. They are expected to be used in MEMS/NEMS.

EELS and Raman spectroscopies can be successfully used for chemical characterization of amorphous carbon coatings. The prevailing atomic arrangement in the DLC coatings is amorphous or quasi-amorphous with small diamond (sp^3), graphite (sp^2) and other unidentifiable micro- or nanocrystallites. Most DLC coatings except those produced by filtered cathodic arc contain from a few to about 50 at.% hydrogen. Sometimes hydrogen is deliberately incorporated in the sputtered and ion plated coatings to tailor their properties.

Amorphous carbon coatings deposited by various techniques exhibit different mechanical and tribological properties. The nanoindenter can be successfully used for measurement of hardness, elastic modulus, fracture toughness, and fatigue life. Microscratch and microwear experiments can be performed using either a nanoindenter or an AFM. Thin coatings deposited by filtered cathodic arc, ion beam and ECR-CVD hold a promise for tribological applications. Coatings as thin as 5 nm or even thinner in thickness provide wear protection. Microscratch, microwear, and accelerated wear testing, if simulated properly can be successfully used to screen coating candidates for industrial applications. In the examples shown in this chapter, trends observed in the microscratch, microwear, and accelerated macrofriction wear tests are similar to that found in functional tests.

References

38.1 B. Bhushan, B.K. Gupta: *Handbook of Tribology: Materials, Coatings, and Surface Treatments* (Krieger, Malabar 1997)

38.2 B. Bhushan: *Principles and Applications of Tribology* (Wiley, New York 1999)

38.3 B. Bhushan: *Introduction to Tribology* (Wiley, New York 2002)

38.4 B. Bhushan, B.K. Gupta, G.W. VanCleef, C. Capp, J.V. Coe: Fullerene (C_{60}) films for solid lubrication, Tribol. Trans. **36**, 573–580 (1993)

38.5 B.K. Gupta, B. Bhushan, C. Capp, J.V. Coe: Material characterization and effect of purity and ion implantation on the friction and wear of sublimed fullerene films, J. Mater. Res. **9**, 2823–2838 (1994)

38.6 B.K. Gupta, B. Bhushan: Fullerene particles as an additive to liquid lubricants and greases for low friction and wear, Lubr. Eng. **50**, 524–528 (1994)

38.7 B. Bhushan, V.V. Subramaniam, A. Malshe, B.K. Gupta, J. Ruan: Tribological properties of polished diamond films, J. Appl. Phys. **74**, 4174–4180 (1993)

38.8 B. Bhushan, B.K. Gupta, V.V. Subramaniam: Polishing of diamond films, Diam. Films Technol. **4**, 71–97 (1994)

38.9 P. Sander, U. Kaiser, M. Altebockwinkel, L. Wiedmann, A. Benninghoven, R.E. Sah, P. Koidl: Depth profile analysis of hydrogenated carbon layers on silicon by x-ray photoelectron spectroscopy, auger electron spectroscopy, electron energy-loss spectroscopy, and secondary ion mass spectrometry, J. Vac. Sci. Technol. A **5**, 1470–1473 (1987)

38.10 A. Matthews, S.S. Eskildsen: Engineering applications for diamond-like carbon, Diam. Relat. Mater. **3**, 902–911 (1994)

38.11 A.H. Lettington: Applications of diamond-like carbon thin films, Carbon **36**, 555–560 (1998)

38.12 B.K. Gupta, B. Bhushan: Mechanical and tribological properties of hard carbon coatings for magnetic recording heads, Wear **190**, 110–122 (1995)

38.13 B. Bhushan: *Tribology and Mechanics of Magnetic Storage Devices*, 2nd edn. (Springer, Berlin, Heidelberg 1996)

38.14 B. Bhushan: *Mechanics and Reliability of Flexible Magnetic Media*, 2nd edn. (Springer, Berlin, Heidelberg 2000)

38.15 B. Bhushan, S.T. Patton, R. Sundaram, S. Dey: Pole tip recession studies of hard carbon-coated thin-film tape heads, J. Appl. Phys. **79**, 5916–5918 (1996)

38.16 J. Xu, B. Bhushan: Pole tip recession studies of thin-film rigid disk head sliders II: Effects of air bearing surface and pole tip region designs and carbon coating, Wear **219**, 30–41 (1998)

38.17 W.W. Scott, B. Bhushan: Corrosion and wear studies of uncoated and ultra-thin DLC coated magnetic tape-write heads and magnetic tapes, Wear **243**, 31–42 (2000)

38.18 W.W. Scott, B. Bhushan: Loose debris and head stain generation and pole tip recession in modern tape drives, J. Inf. Storage Proc. Syst. **2**, 221–254 (2000)

38.19 W.W. Scott, B. Bhushan, A.V. Lakshmikumaran: Ultrathin diamond-like carbon coatings used for reduction of pole tip recession in magnetic tape heads, J. Appl. Phys. **87**, 6182–6184 (2000)

38.20 B. Bhushan: Macro- and microtribology of magnetic storage devices. In: *Modern Tribology Handbook*, ed. by B. Bhushan (CRC, Boca Raton 2001) pp. 1413–1513

38.21 B. Bhushan: Nanotribology and nanomechanics of MEMS devices, Proc. 9th Annu. Workshop Micro Electro Mech. Syst. (IEEE, New York 1996) pp. 91–98

38.22 B. Bhushan (Ed.): *Tribology Issues and Opportunities in MEMS* (Kluwer, Dordrecht 1998)

38.23 B. Bhushan: *Handbook of Micro/Nanotribology*, 2nd edn. (CRC, Boca Raton 1999)

38.24 B. Bhushan: Macro- and microtribology of MEMS materials. In: *Modern Tribology Handbook*, ed. by B. Bhushan (CRC, Boca Raton 2001) pp. 1515–1548

38.25 S. Aisenberg, R. Chabot: Ion beam deposition of thin films of diamond-like carbon, J. Appl. Phys. **49**, 2953–2958 (1971)

38.26 E.G. Spencer, P.H. Schmidt, D.C. Joy, F.J. Sansalone: Ion beam deposited polycrystalline diamond-like films, Appl. Phys. Lett. **29**, 118–120 (1976)

38.27 A. Grill, B.S. Meyerson: Development and status of diamondlike carbon. In: *Synthetic Diamond: Emerging CVD Science and Technology*, ed. by K.E. Spear, J.P. Dismukes (Wiley, New York 1994) pp. 91–141

38.28 Y. Catherine: Preparation techniques for diamond-like carbon. In: *Diamond and Diamond-Like Films and Coatings*, ed. by R.E. Clausing, L.L. Horton, J.C. Angus, P. Koidl (Plenum, New York 1991) pp. 193–227

38.29 J.J. Cuomo, D.L. Pappas, J. Bruley, J.P. Doyle, K.L. Seagner: Vapor deposition processes for amorphous carbon films with sp^3 fractions approaching diamond, J. Appl. Phys. **70**, 1706–1711 (1991)

38.30 J.C. Angus, C.C. Hayman: Low pressure metastable growth of diamond and diamondlike phase, Science **241**, 913–921 (1988)

38.31 J.C. Angus, F. Jensen: Dense diamondlike hydrocarbons as random covalent networks, J. Vac. Sci. Technol. A **6**, 1778–1782 (1988)

38.32 D.C. Green, D.R. McKenzie, P.B. Lukins: The microstructure of carbon thin films, Mater. Sci. Forum **52/53**, 103–124 (1989)

38.33 N.H. Cho, K.M. Krishnan, D.K. Veirs, M.D. Rubin, C.B. Hopper, B. Bhushan, D.B. Bogy: Chemical structure and physical properties of diamond-like amorphous carbon films prepared by magnetron sputtering, J. Mater. Res. **5**, 2543–2554 (1990)

38.34 J.C. Angus: Diamond and diamondlike films, Thin Solid Films **216**, 126–133 (1992)

38.35 B. Bhushan, A.J. Kellock, N.H. Cho, J.W. Ager III: Characterization of chemical bonding and physical characteristics of diamond-like amorphous carbon and diamond films, J. Mater. Res. **7**, 404–410 (1992)

38.36 J. Robertson: Properties of diamond-like carbon, Surf. Coat. Technol. **50**, 185–203 (1992)

38.37 N. Savvides, B. Window: Diamondlike amorphous carbon films prepared by magnetron sputtering of graphite, J. Vac. Sci. Technol. A **3**, 2386–2390 (1985)

38.38 J.C. Angus, P. Koidl, S. Domitz: Carbon thin films. In: *Plasma Deposited Thin Films*, ed. by J. Mort, F. Jensen (CRC, Boca Raton 1986) pp. 89–127

38.39 J. Robertson: Amorphous carbon, Adv. Phys. **35**, 317–374 (1986)

38.40 M. Rubin, C.B. Hooper, N.H. Cho, B. Bhushan: Optical and mechanical properties of DC sputtered carbon films, J. Mater. Res. **5**, 2538–2542 (1990)

38.41 G.J. Vandentop, M. Kawasaki, R.M. Nix, I.G. Brown, M. Salmeron, G.A. Somorjai: Formation of hydrogenated amorphous carbon films of controlled hardness from a methane plasma, Phys. Rev. B **41**, 3200–3210 (1990)

38.42 J.J. Cuomo, D.L. Pappas, R. Lossy, J.P. Doyle, J. Bruley, G.W. Di Bello, W. Krakow: Energetic carbon deposition at oblique angles, J. Vac. Sci. Technol. A **10**, 3414–3418 (1992)

38.43 D.L. Pappas, K.L. Saenger, J. Bruley, W. Krakow, J.J. Cuomo: Pulsed laser deposition of diamondlike carbon films, J. Appl. Phys. **71**, 5675–5684 (1992)

38.44 H.J. Scheibe, B. Schultrich: DLC film deposition by laser-arc and study of properties, Thin Solid Films **246**, 92–102 (1994)

38.45 C. Donnet, A. Grill: Friction control of diamond-like carbon coatings, Surf. Coat. Technol. **94/95**, 456 (1997)

38.46 A. Grill: Tribological properties of diamondlike carbon and related materials, Surf. Coat. Technol. **94/95**, 507 (1997)

38.47 B.K. Gupta, B. Bhushan: Micromechanical properties of amorphous carbon coatings deposited by different deposition techniques, Thin Solid Films **270**, 391–398 (1995)

38.48 X. Li, B. Bhushan: Micro/nanomechanical and tribological characterization of ultra-thin amorphous carbon coatings, J. Mater. Res. **14**, 2328–2337 (1999)

38.49 X. Li, B. Bhushan: Mechanical and tribological studies of ultra-thin hard carbon overcoats for magnetic recording heads, Z. Metallkd. **90**, 820–830 (1999)

38.50 S. Sundararajan, B. Bhushan: Micro/nanotribology of ultra-thin hard amorphous carbon coatings using atomic force/friction force microscopy, Wear **225–229**, 678–689 (1999)

38.51 B. Bhushan: Chemical, mechanical, and tribological characterization of ultra-thin and hard amorphous carbon coatings as thin as 3.5 nm: Recent developments, Diam. Relat. Mater. **8**, 1985–2015 (1999)

38.52 I.I. Aksenov, V.E. Strel'Nitskii: Wear resistance of diamond-like carbon coatings, Surf. Coat. Technol. **47**, 252–256 (1991)

38.53 D.R. McKenzie, D. Muller, B.A. Pailthorpe, Z.H. Wang, E. Kravtchinskaia, D. Segal, P.B. Lukins, P.J. Martin, G. Amaratunga, P.H. Gaskell, A. Saeed: Properties of tetrahedral amorphous carbon prepared by vacuum arc deposition, Diam. Relat. Mater. **1**, 51–59 (1991)

38.54 R. Lossy, D.L. Pappas, R.A. Roy, J.J. Cuomo: Filtered arc deposition of amorphous diamond, Appl. Phys. Lett. **61**, 171–173 (1992)

38.55 I.G. Brown, A. Anders, S. Anders, M.R. Dickinson, I.C. Ivanov, R.A. MacGill, X.Y. Yao, K.M. Yu: Plasma synthesis of metallic and composite thin films with atomically mixed substrate bonding, Nucl. Instrum. Meth. B **80/81**, 1281–1287 (1993)

38.56 P.J. Fallon, V.S. Veerasamy, C.A. Davis, J. Robertson, G.A.J. Amaratunga, W.I. Milne, J. Koskinen: Properties of filtered-ion-beam-deposited diamond-like carbon as a function of ion energy, Phys. Rev. B **48**, 4777–4782 (1993)

38.57 S. Anders, A. Anders, I.G. Brown, B. Wei, K. Komvopoulos, J.W. Ager III, K.M. Yu: Effect of vacuum arc deposition parameters on the properties of amorphous carbon thin films, Surf. Coat. Technol. **68/69**, 388–393 (1994)

38.58 S. Anders, A. Anders, I.G. Brown, M.R. Dickinson, R.A. MacGill: Metal plasma immersion ion implantation and deposition using arc plasma sources, J. Vac. Sci. Technol. B **12**, 815–820 (1994)

38.59　S. Anders, A. Anders, I.G. Brown: Transport of vacuum arc plasma through magnetic macroparticle filters, Plasma Sources Sci. **4**, 1–12 (1995)

38.60　D.M. Swec, M.J. Mirtich, B.A. Banks: *Ion Beam and Plasma Methods of Producing Diamondlike Carbon Films* (NASA, Cleveland 1989), Report No. NASA TM102301

38.61　A. Erdemir, M. Switala, R. Wei, P. Wilbur: A tribological investigation of the graphite-to-diamond-like behavior of amorphous carbon films ion beam deposited on ceramic substrates, Surf. Coat. Technol. **50**, 17–23 (1991)

38.62　A. Erdemir, F.A. Nicols, X.Z. Pan, R. Wei, P.J. Wilbur: Friction and wear performance of ion-beam deposited diamond-like carbon films on steel substrates, Diam. Relat. Mater. **3**, 119–125 (1993)

38.63　R. Wei, P.J. Wilbur, M.J. Liston: Effects of diamond-like hydrocarbon films on rolling contact fatigue of bearing steels, Diam. Relat. Mater. **2**, 898–903 (1993)

38.64　A. Erdemir, C. Donnet: Tribology of diamond, diamond-like carbon, and related films. In: *Modern Tribology Handbook*, ed. by B. Bhushan (CRC, Boca Raton 2001) pp. 871–908

38.65　J. Asmussen: Electron cyclotron resonance microwave discharges for etching and thin-film deposition, J. Vac. Sci. Technol. A **7**, 883–893 (1989)

38.66　J. Suzuki, S. Okada: Deposition of diamondlike carbon films using electron cyclotron resonance plasma chemical vapor deposition from ethylene gas, Jpn. J. Appl. Phys. **34**, L1218–L1220 (1995)

38.67　B.A. Banks, S.K. Rutledge: Ion beam sputter deposited diamond like films, J. Vac. Sci. Technol. **21**, 807–814 (1982)

38.68　C. Weissmantel, K. Bewilogua, K. Breuer, D. Dietrich, U. Ebersbach, H.J. Erler, B. Rau, G. Reisse: Preparation and properties of hard i-C and i-BN coatings, Thin Solid Films **96**, 31–44 (1982)

38.69　H. Dimigen, H. Hubsch: Applying low-friction wear-resistant thin solid films by physical vapor deposition, Philips Tech. Rev. **41**, 186–197 (1983)

38.70　B. Bhushan, B.K. Gupta, M.H. Azarian: Nanoindentation, microscratch, friction and wear studies for contact recording applications, Wear **181–183**, 743–758 (1995)

38.71　B. Bhushan, L. Yang, C. Gao, S. Suri, R.A. Miller, B. Marchon: Friction and wear studies of magnetic thin-film rigid disks with glass-ceramic, glass and aluminum-magnesium substrates, Wear **190**, 44–59 (1995)

38.72　L. Holland, S.M. Ojha: Deposition of hard and insulating carbonaceous films of an RF target in butane plasma, Thin Solid Films **38**, L17–L19 (1976)

38.73　L.P. Andersson: A review of recent work on hard i-C films, Thin Solid Films **86**, 193–200 (1981)

38.74　A. Bubenzer, B. Dischler, B. Brandt, P. Koidl: RF plasma deposited amorphous hydrogenated hard carbon thin films, preparation, properties and applications, J. Appl. Phys. **54**, 4590–4594 (1983)

38.75　A. Grill, B.S. Meyerson, V.V. Patel: Diamond-like carbon films by RF plasma-assisted chemical vapor deposition from acetylene, IBM J. Res. Dev. **34**, 849–857 (1990)

38.76　A. Grill, B.S. Meyerson, V.V. Patel: Interface modification for improving the adhesion of a-C:H to metals, J. Mater. Res. **3**, 214 (1988)

38.77　A. Grill, V.V. Patel, B.S. Meyerson: Optical and tribological properties of heat-treated diamond-like carbon, J. Mater. Res. **5**, 2531–2537 (1990)

38.78　F. Jansen, M. Machonkin, S. Kaplan, S. Hark: The effect of hydrogenation on the properties of ion beam sputter deposited amorphous carbon, J. Vac. Sci. Technol. A **3**, 605–609 (1985)

38.79　S. Kaplan, F. Jansen, M. Machonkin: Characterization of amorphous carbon-hydrogen films by solid-state nuclear magnetic resonance, Appl. Phys. Lett. **47**, 750–753 (1985)

38.80　H.C. Tsai, D.B. Bogy, M.K. Kundmann, D.K. Veirs, M.R. Hilton, S.T. Mayer: Structure and properties of sputtered carbon overcoats on rigid magnetic media disks, J. Vac. Sci. Technol. A **6**, 2307–2315 (1988)

38.81　B. Marchon, M. Salmeron, W. Siekhaus: Observation of graphitic and amorphous structures on the surface of hard carbon films by scanning tunneling microscopy, Phys. Rev. B **39**, 12907–12910 (1989)

38.82　B. Dischler, A. Bubenzer, P. Koidl: Hard carbon coatings with low optical absorption, Appl. Phys. Lett. **42**, 636–638 (1983)

38.83　D.S. Knight, W.B. White: Characterization of diamond films by Raman spectroscopy, J. Mater. Res. **4**, 385–393 (1989)

38.84　J.W. Ager III, D.K. Veirs, C.M. Rosenblatt: Spatially resolved Raman studies of diamond films grown by chemical vapor deposition, Phys. Rev. B **43**, 6491–6499 (1991)

38.85　W. Scharff, K. Hammer, O. Stenzel, J. Ullman, M. Vogel, T. Frauenheim, B. Eibisch, S. Roth, S. Schulze, I. Muhling: Preparation of amorphous i-C films by ion-assisted methods, Thin Solid Films **171**, 157–169 (1989)

38.86　B. Bhushan, X. Li: Nanomechanical characterization of solid surfaces and thin films, Int. Mater. Rev. **48**, 125–164 (2003)

38.87　X. Li, D. Diao, B. Bhushan: Fracture mechanisms of thin amorphous carbon films in nanoindentation, Acta Mater. **45**, 4453–4461 (1997)

38.88　X. Li, B. Bhushan: Measurement of fracture toughness of ultra-thin amorphous carbon films, Thin Solid Films **315**, 214–221 (1998)

38.89　X. Li, B. Bhushan: Evaluation of fracture toughness of ultra-thin amorphous carbon coatings deposited by different deposition techniques, Thin Solid Films **355/356**, 330–336 (1999)

38.90 X. Li, B. Bhushan: Development of a nanoscale fatigue measurement technique and its application to ultrathin amorphous carbon coatings, Scr. Mater. **47**, 473–479 (2002)

38.91 X. Li, B. Bhushan: Nanofatigue studies of ultrathin hard carbon overcoats used in magnetic storage devices, J. Appl. Phys. **91**, 8334–8336 (2002)

38.92 J. Robertson: Deposition of diamond-like carbon, Philos. Trans. R. Soc. Lond. A **342**, 277–286 (1993)

38.93 S.J. Bull: Tribology of carbon coatings: DLC, diamond and beyond, Diam. Relat. Mater. **4**, 827–836 (1995)

38.94 N. Savvides, T.J. Bell: Microhardness and Young's modulus of diamond and diamondlike carbon films, J. Appl. Phys. **72**, 2791–2796 (1992)

38.95 B. Bhushan, M.F. Doerner: Role of mechanical properties and surface texture in the real area of contact of magnetic rigid disks, ASME J. Tribol. **111**, 452–458 (1989)

38.96 S. Suresh: *Fatigue of Materials* (Cambridge Univ. Press, Cambridge 1991)

38.97 D.B. Marshall, A.G. Evans: Measurement of adherence of residual stresses in thin films by indentation. I. Mechanics of interface delamination, J. Appl. Phys. **15**, 2632–2638 (1984)

38.98 A.G. Evans, J.W. Hutchinson: On the mechanics of delamination and spalling in compressed films, Int. J. Solids Struct. **20**, 455–466 (1984)

38.99 S. Sundararajan, B. Bhushan: Development of a continuous microscratch technique in an atomic force microscope and its application to study scratch resistance of ultrathin hard amorphous carbon coatings, J. Mater. Res. **16**, 437–445 (2001)

38.100 B. Bhushan, B.K. Gupta: Micromechanical characterization of Ni-P coated aluminum-magnesium, glass and glass-ceramic substrates and finished magnetic thin-film rigid disks, Adv. Inf. Storage Syst. **6**, 193–208 (1995)

38.101 X. Li, B. Bhushan: Micromechanical and tribological characterization of hard amorphous carbon coatings as thin as 5 nm for magnetic recording heads, Wear **220**, 51–58 (1998)

38.102 B. Bhushan, V.N. Koinkar: Microscale mechanical and tribological characterization of hard amorphous coatings as thin as 5 nm for magnetic disks, Surf. Coat. Technol. **76/77**, 655–669 (1995)

38.103 V.N. Koinkar, B. Bhushan: Microtribological properties of hard amorphous carbon protective coatings for thin-film magnetic disks and heads, Proc. Inst. Mech. Eng. J **211**, 365–372 (1997)

38.104 T.W. Wu: Microscratch and load relaxation tests for ultra-thin films, J. Mater. Res. **6**, 407–426 (1991)

38.105 R. Memming, H.J. Tolle, P.E. Wierenga: Properties of polymeric layers of hydrogenated amorphous carbon produced by plasma-activated chemical vapor deposition: tribological and mechanical properties, Thin Solid Films **143**, 31–41 (1986)

38.106 C. Donnet, T. Le Mogne, L. Ponsonnet, M. Belin, A. Grill, V. Patel: The respective role of oxygen and water vapor on the tribology of hydrogenated diamond-like carbon coatings, Tribol. Lett. **4**, 259 (1998)

38.107 F.P. Bowden, J.E. Young: Friction of diamond, graphite and carbon and the influence of surface films, Proc. R. Soc. Lond. A **208**, 444–455 (1951)

38.108 B. Marchon, N. Heiman, M.R. Khan: Evidence for tribochemical wear on amorphous carbon thin films, IEEE Trans. Magn. **26**, 168–170 (1990)

38.109 M.T. Dugger, Y.W. Chung, B. Bhushan, W. Rothschild: Friction, wear, and interfacial chemistry in thin film magnetic rigid disk files, ASME J. Tribol. **112**, 238–245 (1990)

38.110 B.D. Strom, D.B. Bogy, C.S. Bhatia, B. Bhushan: Tribochemical effects of various gases and water vapor on thin film magnetic disks with carbon overcoats, ASME J. Tribol. **113**, 689–693 (1991)

38.111 B. Bhushan, J. Ruan: Tribological performance of thin film amorphous carbon overcoats for magnetic recording rigid disks in various environments, Surf. Coat. Technol. **68/69**, 644–650 (1994)

38.112 B. Bhushan, G.S.A.M. Theunissen, X. Li: Tribological studies of chromium oxide films for magnetic recording applications, Thin Solid Films **311**, 67–80 (1997)

39. Self-Assembled Monolayers for Nanotribology and Surface Protection

Bharat Bhushan

Reliability of various micro- and nanodevices requiring relative motion as well as magnetic storage devices requires the use of hydrophobic and lubricating films to minimize adhesion, stiction, friction, and wear. In various applications, surfaces need to be protected from exposure to the operating environment, and hydrophobic films are of interest. The surface films should be molecularly thick, well-organized, chemically bonded to the substrate, and insensitive to environment. Ordered molecular assemblies with high hydrophobicity can be engineered using chemical grafting of various polymer molecules with suitable functional head groups, spacer chains, and nonpolar surface terminal groups.

In this chapter, we focus on self-assembled monolayers (SAMs) with high hydrophobicity and good nanotribological properties. SAMs are produced by various organic precursors. We first present a primer to organic chemistry, followed by an overview of selected SAMs with various substrates, spacer chains, and terminal groups in the molecular chains and an overview of nanotribological properties of SAMs. The contact angle, adhesion, friction, and wear properties of SAMs having various spacer chains with different surface terminal and head groups (hexadecane thiol, biphenyl thiol, perfluoroalkylsilane, alkylsilane, perfluoroalkylphosphonate, and alkylphosphonate) on various substrates (Au, Si, and Al) are surveyed. Chemical degradation mechanisms and environmental effects are studied. Based on the contact angle and nanotribological properties of

39.1	**Background**	1309
	39.1.1 Need for Hydrophobic Surfaces for Nanotribology	1310
	39.1.2 Surface Films for Nanotribology and Surface Protection	1310
	39.1.3 Scope of the Chapter	1312
39.2	**A Primer to Organic Chemistry**	1313
	39.2.1 Electronegativity/Polarity	1313
	39.2.2 Classification and Structure of Organic Compounds	1314
	39.2.3 Polar and Nonpolar Groups	1316
39.3	**Self-Assembled Monolayers: Substrates, Spacer Chains, and End Groups in the Molecular Chains**	1316
39.4	**Contact Angle and Nanotribological Properties of SAMs**	1319
	39.4.1 Measurement Techniques	1322
	39.4.2 Hexadecane Thiol and Biphenyl Thiol SAMs on Au(111)	1323
	39.4.3 Perfluoroalkylsilane and Alkylsilane SAMs on Si(100) and Perfluoroalkylphosphonate and Alkylphosphonate SAMS on Al	1329
	39.4.4 Chemical Degradation and Environmental Studies	1338
39.5	**Summary**	1340
	References	1342

various SAM films by atomic force microscopy (AFM) it is found that perfluoroalkylsilane and perfluorophosphonate SAMs exhibit attractive hydrophobic and tribological properties.

39.1 Background

Reliability of various micro- and nanodevices, also commonly referred to as micro-/nanoelectromechanical systems (MEMS/NEMS) and bioMEMS/bioNEMS, requiring relative motion, as well as magnetic storage devices (which include magnetic rigid disk and tape drives) requires the use of hydrophobic and lubricating films to minimize adhesion, stiction, friction, and wear [39.1–10]. In various applications, surfaces need to be protected from exposure to the operating environment. For example, in various biomedical applications, such as biosensors and implantable biomedical devices, undesirable protein adsorption, biofouling, and biocompatibility are some of the major issues [39.5, 7]. In micro- and nanofluidic-based sensors, the fluid drag in micro- and nanochannels can be reduced by using hydrophobic coatings. Selected hydrophobic films are needed for these applications.

39.1.1 Need for Hydrophobic Surfaces for Nanotribology

The source of the liquid film at the interface can be a preexisting film of liquid and/or capillary condensates of water vapor from the environment. If the liquid wets the surface ($0 \leq \theta < 90°$, where θ is the contact angle between the liquid–vapor interface and the liquid–solid interface for a liquid droplet sitting on a solid surface (Fig. 39.1a)), the liquid surface is constrained to lie parallel to the surface [39.11–13], and the complete liquid surface is therefore concave in shape (Fig. 39.1b). Direct measurement of contact angle is most widely made from sessile drops. The angle is generally measured by aligning a tangent with the drop profile at the point of contact with the solid surface using a telescope equipped with a goniometer eyepiece. Surface tension results in a pressure difference across any meniscus surface, referred to as capillary pressure or Laplace pressure, which is negative for a concave meniscus [39.14, 15]. The negative Laplace pressure results in an intrinsic attractive (adhesive) force which depends on the interface roughness (the local geometry of interacting asperities and number of asperities), the surface tension, and the contact angle. During normal separation, this intrinsic force needs to be overcome [39.16]. During sliding, frictional effects need to be overcome, not only because of external load but also because of intrinsic adhesive force. A measured value of high static friction force contributed largely by liquid-mediated adhesion (meniscus contribution) is generally referred to as *stiction*. It becomes a major concern in micro- and nanodevices operating at ultralow loads, as the liquid-mediated adhesive force may be on the same order as the external load. The effect of liquid-mediated adhesion can be minimized by increasing surface roughness and/or the use of a liquid with low surface tension with film thickness on the order of the surface roughness, as well as its chemical bonding to the substrate [39.14–21]. The formation of menisci and/or condensation of water vapor from the environment at the interface can be minimized by the use of hydrophobic (water-fearing) coatings. Surfaces can be made superhydrophobic by the introduction of controlled roughness on the surfaces to take advantage of the so-called lotus effect [39.22, 23].

39.1.2 Surface Films for Nanotribology and Surface Protection

Surfaces can be treated or coated with a liquid with relatively low surface tension or certain solid films to make them hydrophobic and/or to control adhesion, stiction, friction, and wear.

The classic approach to lubrication uses freely supported multimolecular layers of liquid lubricants [39.2, 4, 6, 14, 15, 20, 24–27]. Boundary lubricant films are formed by physisorption, chemisorption or chemical reaction. The physisorbed films can be either monomolecularly or polymolecularly thick. Chemisorbed films are monomolecular, but stoichiometric films formed by chemical reaction can be multilayers. In general, the stability and durability of surface films decrease in the following order: chemically reacted films > chemisorbed films > physisorbed films. A good boundary lubricant

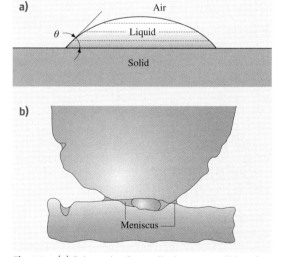

Fig. 39.1 (a) Schematic of a sessile drop on a solid surface and the definition of contact angle. (b) Formation of meniscus bridges as a result of liquid present at an interface

should have a high degree of interaction between its molecules and the sliding surface. As a general rule, liquids will have a more desirable performance when they are polar and thus able to grip onto solid surfaces (or be adsorbed). Polar lubricants contain reactive functional end groups. Boundary lubrication properties are also dependent upon the molecular conformation and lubricant spreading. It should be noted that liquid films with thickness on the order of a few nm may be discontinuous and may deposit in island form with nonuniform thickness and lateral resolution on the nm scale.

Solid films are also commonly used for controlling hydrophobicity and/or adhesion, stiction, friction, and wear. Hydrophobic films have nonpolar surface terminal groups (to be described later) which repel water. These films have low surface energy (15–30 dyn/cm) and high contact angle ($\theta \geq 90°$) which minimize wetting (e.g., [39.25, 28, 29]). Multimolecularly thick (few tenths of nm) films of conventional solid lubricants have been studied. *Hansma* et al. [39.30] reported the deposition of multimolecularly thick, highly oriented polytetrafluoroethylene (PTFE) films from the melt or vapor phase or from solution by a mechanical deposition technique by dragging the polymer at controlled temperature, pressure, and speed against a smooth glass substrate. *Scandella* et al. [39.31] reported that the coefficient of nanoscale friction of MoS_2 platelets on mica, obtained by the exfoliation of lithium intercalated MoS_2 in water, was a factor of 1.4 less than that of mica itself. However, MoS_2 is reactive to water, and its friction and wear properties degrade with increasing humidity [39.14, 15]. Amorphous diamond-like carbon (DLC) coatings can be produced with extremely high hardness and are used commercially as wear-resistant coatings [39.32, 33]. They are widely used in magnetic storage devices [39.2]. Doping of the DLC matrix with elements such as hydrogen, nitrogen, oxygen, silicon, and fluorine influences their hydrophobicity and tribological properties [39.32, 34, 35]. Nitrogen and oxygen reduce the contact angle (or increase the surface energy) due to the strong polarity formed when these elements bond to carbon. On the other hand, silicon and fluorine increase the contact angle to 70–100° (or reduce the surface energy to 20–40 dyn/cm), making them hydrophobic [39.36, 37]. Nanocomposite coatings with a diamond-like carbon (a-C:H) network and a glasslike a-Si:O network are generally deposited using a plasma-enhanced chemical vapor deposition (PECVD) technique in which plasma is formed from a siloxane precursor using a hot filament. For fluorinated DLC, CF_4 is added as the fluorocarbon source to an acetylene plasma. In addition, fluorination of DLC can be achieved by postdeposition treatment of DLC coatings in CF_4 plasma. Silicon- and fluorine-containing DLC coatings mainly reduce their polarity due to the loss of sp^2 bonded carbon (due to the polarization potential of the involved π electrons) and dangling bonds of the DLC network. As silicon and fluorine are unable to form double bonds, they force carbon into a sp^3 bonding state [39.37]. Friction and wear properties of both silicon-containing and fluorinated DLC coatings have been reported to be superior to those of conventional DLC coatings [39.38, 39]. However, DLC coatings require a line-of-sight deposition process which prevents deposition on complex geometries. Furthermore, it has been reported that some self-assembled monolayers (SAMs) are superior to DLC coatings in terms of their hydrophobicity and tribological performance [39.40, 41].

Organized and dense molecular-scale layers of, preferably long-chain, organic molecules are known to be superior lubricants on both macro- and micro-/nanoscales as compared with freely supported multi-molecular layers [39.4, 6]. Common techniques to produce molecular scale organized layers are Langmuir–Blodgett (LB) deposition and chemical grafting of organic molecules to realize SAMs [39.28, 29]. In the LB technique, organic molecules from suitable amphiphilic molecules are first organized at the air–water interface and then physisorbed on a solid surface to form mono- or multimolecular layers [39.42]. in the case of SAMs the functional groups of molecules chemisorb onto a solid surface, which results in the spontaneous formation of robust, highly ordered, oriented, dense monolayers [39.29]. In both cases, the organic molecules used have well-distinguished amphiphilic properties (a hydrophilic functional head and a hydrophobic aliphatic tail) so that adsorption of such molecules on an active inorganic substrate leads to their firm attachment to the surface. Direct organization of SAMs on the solid surfaces allows coating in inaccessible areas. The weak adhesion of classical LB films to the substrate surface restricts their lifetime during sliding, whereas certain SAMs can be very durable. As a result, SAMs are of great interest in tribological applications.

Much research into the application of SAMs has been carried out using the so-called soft lithographic technique [39.43, 44]. This is a nonphotolithographic technique. Photolithography is based on a projection-printing system used for projection of an image from a mask to a thin-film photoresist; its resolution is limited by optical diffraction limits. In soft lithography, an elastomeric stamp or mold is used to generate micropatterns of SAMs by either contact printing (known as microcontact printing, μCP [39.45]), by embossing (nanoimprint lithography) [39.46] or by replica molding [39.47], thereby circumventing the diffraction limits of photolithography. The stamps are generally cast from photolithographically generated patterned masters, and the stamp material is generally polydimethylsiloxane (PDMS). In μCP, the ink is a SAM precursor to produce nm-thick resists with lines thinner than 100 nm. Soft lithography requires little capital investment. μCP and embossing techniques may be used to produce microdevices which are substantially cheaper and more flexible in terms of the choice of material for construction than with conventional photolithography (e.g., SAMs and non-SAM entities for μCP and elastomers for embossing).

The largest industrial application for SAMs is in digital micromirror devices (DMD) used in optical projection displays [39.48, 49]. The chip set of a DMD consists of half a million to more than two million independently controlled reflective aluminum alloy micromirrors of about $12\,\mu m^2$. These micromirrors switch forward and backward at a frequency on the order of 5–7 kHz with a rotation of $\pm 12°$ with respect to the horizontal plane; the movement is limited by a mechanical stop. Mechanical contact leads to stiction and wear in contacting surfaces. A SAM of vapor-deposited perfluorinated n-alkanoic acid ($C_n F_{2n-1} O_2 H$) [e.g., perfluorodecanoic acid (PFDA), $CF_3(CF_2)_8 COOH$] is used to coat contacting surfaces to make them hydrophobic in order to minimize meniscus formation. Furthermore, the entire DMD chip set is hermetically sealed in order to prevent particulate contamination and excessive condensation of water at the contacting surfaces. A so-called *getter* strip of PFDA is included inside the hermetically sealed enclosure containing the chip, which acts as a reservoir in order to maintain a PFDA vapor within the package. Degradation mechanisms of SAMs leading to stiction have been studied by *Liu* and *Bhushan* [39.50, 51]. Nanotribological studies of various SAMs on Al substrates have been carried out by *Tambe* and *Bhushan* [39.52], *Bhushan* et al. [39.53], *Hoque* et al. [39.54–56], and *DeRose* et al. [39.57].

There are various other micro-/nanodevices which require SAMs for hydrophobicity in order to minimize meniscus formation. Examples include micromotors, microgears, microvalves, microswitches, mirror-based optical switches, and atomic force microscopy probes [39.5–7]. Nanotribological studies on Si substrates have been carried out by *Bhushan* et al. [39.58–60], *Kasai* et al. [39.61], *Lee* et al. [39.62], *Tambe* and *Bhushan* [39.52], and *Tao* and *Bhushan* [39.63, 64]. SAM deposition on Cu surfaces is also being explored for corrosion inhibition for micro-/ nanoelectronics and/or heat-exchange surfaces, exploiting dropwise condensation [39.65–67].

Other industrial applications for SAMs are in the areas of biochemical and optical sensors, devices for use as drug-delivery vehicles, and in the construction of electronic components [39.68–71]. Biochemical sensors require highly sensitive organic layers with tailored biological properties that can be incorporated into electronic, optical or electrochemical devices. Self-assembled microscopic vesicles are being developed to ferry potentially life-saving drugs to cancer patients. By assembling organic, metal, and phosphonate molecules (complexes of phosphorous and oxygen atoms) into conductive materials, these can be produced as self-made sandwiches for use as electronic components. Several applications have been proposed based on silicon, glass or polymer nanochannels, including cell immunoisolation chambers, DNA separation devices, and biocapsules for drug delivery [39.5, 6].

SAMs are also being considered for protection of surfaces from exposure to the operating environment. They are being developed to reduce corrosion and oxidation of Cu in heat exchangers [39.65–67]. They are being developed to minimize undesirable protein adsorption and biofouling, and improve biocompatibility in biosensors and implantable biomedical devices [39.52, 53, 58–63, 72, 73]. These films can also be used to reduce fluid drag in micro-/ nanochannels.

39.1.3 Scope of the Chapter

An overview of molecularly thick layers of liquid lubricants and conventional solid lubricants can be found in various references, such as works by *Bhushan* [39.2, 4, 6, 14, 15, 18, 32], *Bhushan* and *Zhao* [39.20], and *Liu* [39.27]. In this chapter, we focus on SAMs for high hydrophobicity, and low adhesion, friction, and wear. SAMs are produced by various organic precursors. We first present a primer to organic chemistry followed by an overview on suitable substrates, spacer chains, and end groups in molecular chains, an overview on contact angle, adhesion, friction, and wear properties of various SAMs, and some concluding remarks.

39.2 A Primer to Organic Chemistry

All organic compounds contain the carbon (C) atom. Carbon, in combination with hydrogen, oxygen, nitrogen, sulfur, and phosphor, results in a large number of organic compounds. The atomic number of carbon is 6, and its electron structure is $1s^2\,2s^2\,2p^2$. Two stable isotopes of carbon, ^{12}C and ^{13}C, exist. With four electrons in its outer shell, carbon forms four covalent bonds, with each bond resulting from two atoms sharing a pair of electrons. The number of electron pairs that two atoms share determines whether or not the bond is single or multiple. In a single bond, only one pair of electrons is shared by the atoms. Carbon can also form multiple bonds by sharing two or three pairs of electrons between the atoms. For example, the double bond formed by sharing two electron pairs is stronger than a single bond, and it is shorter than a single bond. An organic compound is classified as saturated if it contains only single bonds and as unsaturated if the molecules possess one or more multiple carbon–carbon bonds.

39.2.1 Electronegativity/Polarity

When two different kinds of atoms share a pair of electrons, a bond is formed in which electrons are shared unequally; one atom assumes a partial positive charge and the other a negative charge with respect to each other. This difference in charge occurs because the two atoms exert unequal attraction on the pair of shared electrons. The attractive force that an atom of an element has for shared electrons in a molecule or polyatomic ion is known as its electronegativity. Elements differ in their electronegativities. A scale of relative electronegatives, in which the most electronegative element, fluorine, is assigned a value of 4.0, was developed by Pauling. Relative electronegativities of the elements in the Periodic Table can be found in most undergraduate chemistry textbooks [39.74]. The relative electronegativity of nonmetals is high compared with that of metals. The relative electronegativity of selected elements of interest with high values is presented in Table 39.1.

The polarity of a bond is determined by the difference in electronegativity values of the atoms forming the bond. If the electronegativities are the same the bond is nonpolar, and the electrons are shared equally. In this type of bond, there is no separation of positive and negative charge between atoms. If the atoms have greatly different electronegativities the bond is very polar. A dipole is a molecule that is electrically asymmetrical, causing it to be oppositely charged at two points. As an example, in HCl both hydrogen and chlorine need one electron to form stable electron configurations. They share a pair of electrons. Chlorine is more electronegative and therefore has a greater attraction for the shared electrons than does hydrogen. As a result, the pair of electrons is displaced towards the chlorine atom, giving it a partial negative charge and leaving the hydrogen atom with a partial positive charge (Fig. 39.2). However, the entire HCl molecule is electrically neutral. The hydrogen atom with a partial positive charge (exposed proton on one end) can be easily attracted to the negative charge of other molecules and this is responsible for the polarity of the molecule. A partial charge is usually indicated by δ, and the electronic structure of HCl is given as

$$\overset{\delta+}{\text{H}}\;\;\overset{\delta-}{:\ddot{\text{Cl}}:\,\cdot}$$

Table 39.1 Relative electronegativity of selected elements

Element	Relative electronegativity
F	4.0
O	3.5
N	3.0
Cl	3.0
C	2.5
S	2.5
P	2.1
H	2.1

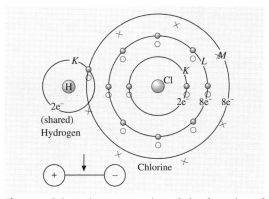

Fig. 39.2 Schematic representation of the formation of a polar HCl molecule

Similar to the HCl molecule, HF is polar, and both behave as a small dipole. On the other hand, methane (CH_4), carbon tetrachloride (CCl_4), and carbon dioxide (CO_2) are nonpolar. In CH_4 and CCl_4, the four C–H and C–Cl polar bonds are identical, and because these bonds emanate from the center to the corners of a tetrahedron in the molecule, the effects of their polarities cancel one another. CO_2 (O=C=O) is nonpolar because the C–O dipoles cancel each other by acting in opposite direction. Water (H–O–H) is a polar molecule. If the atoms in water were linear as in CO_2, the two O–H dipoles would cancel each other, and the molecule would be nonpolar. However, water has a bent structure with an angle of 105° between the two bonds, which is responsible for water being a polar molecule.

39.2.2 Classification and Structure of Organic Compounds

Table 39.2 presents selected organic compounds grouped into classes.

Hydrocarbons

Hydrocarbons are compounds that are composed entirely of carbon and hydrogen atoms bonded to each other by covalent bonds. Saturated hydrocarbons (alkanes) contain single bonds. Unsaturated hydrocarbons that contain C=C bonds are called alkenes, and ones with triple bonds are called alkynes. Unsaturated hydro-

Table 39.2 (a) Names and formulas of selected hydrocarbons

Name	Formula
Saturated hydrocarbons	
Straight-chain alkanes	C_nH_{2n+2}
e.g., methane	CH_4
ethane	C_2H_6 or CH_3CH_3
Alkyl groups	C_nH_{2n+1}
e.g., methyl	$-CH_3$
ethyl	$-CH_2CH_3$
Unsaturated hydrocarbons	
Alkenes	$(CH_2)_n$
e.g., ethene	C_2H_4 or $CH_2=CH_2$
propene	C_3H_6 or $CH_3CH=CH_2$
Alkynes	
e.g., acetylene	$HC\equiv CH$
Aromatic hydrocarbons	
e.g., benzene	C_6H_5OH or ⬡

Table 39.2 (b) Names and formulas of selected alcohols, ethers, phenols, and thiols

Name	Formula
Alcohols	R–OH
e.g., methanol	CH_3OH
ethanol	CH_3CH_2OH
Ethers	R–O–R'
e.g., dimethyl ether	CH_3-O-CH_3
diethyl ether	$CH_3CH_2-O-CH_2CH_3$
Phenols	C_6H_5OH or ⬡–OH
Thiols	–SH
e.g., methanethiol	CH_3SH

The letters R and R' represent an alkyl group. The R-groups in ethers can be the same or different and can be alkyl or aromatic (Ar) groups

carbons that contain aromatic rings, e.g., benzene rings, are called aromatic hydrocarbons.

Saturated Hydrocarbons: Alkanes. The alkanes, also known as paraffins, are saturated hydrocarbons, straight- or branched-chain hydrocarbons with only single covalent bonds between the carbon atoms. The general molecular formula for the alkanes is C_nH_{2n+2}, where n is the number of carbon atoms in the molecule. Each carbon atom is connected to four other atoms by four single covalent bonds. These bonds are separated by angles of 109.5° (the angle between lines from the center of a regular tetrahedron to its corners). Alkane molecules contain only carbon–carbon and carbon–hydrogen bonds, which are symmetrically directed towards the corners of a tetrahedron. Therefore alkane molecules are essentially nonpolar.

Common alkyl groups have the general formula C_nH_{2n+1} (one hydrogen atom fewer than the corresponding alkane). The missing H atom may be detached from any carbon in the alkane. The name of the group is formed from the name of the corresponding alkane by replacing -ane with -yl ending. Some examples are shown in Table 39.2a.

Unsaturated Hydrocarbons. Unsaturated hydrocarbons consist of three families of compounds that contain fewer hydrogen atoms than the alkane with the corresponding number of carbon atoms, and contain multiple bonds between carbon atoms. These include alkenes (with C=C bonds), alkynes (with C≡C bonds), and aromatic compounds (with benzene rings which are

Table 39.2 (c) Names and formulas of selected aldehydes and ketones

Name	Formula
Aldehydes	RCHO or $\underset{\underset{H}{\|}}{R-C=O}$
	ArCHO or $\underset{\underset{H}{\|}}{Ar-C=O}$
e.g., methanal or formaldehyde	HCHO
ethanol or acetaldehyde	CH_3CHO
Ketones	RCOR' or $\underset{\underset{R'}{\|}}{R-C=O}$
	RCOAr or $\underset{\underset{Ar}{\|}}{R-C=O}$
	ArCOAr or $\underset{\underset{Ar}{\|}}{Ar-C=O}$
e.g., butanone or methyl ethyl ketone	$CH_3COCH_2CH_3$

The letters R and R' represent an alkyl group and Ar represents an aromatic group

Table 39.2 (d) Names and formulas of selected carboxylic acids and esters

Name	Formula
Carboxylic acid[a]	RCOOH or $\underset{\underset{OH}{\|}}{R-C=O}$
	ArCOOH or $\underset{\underset{OH}{\|}}{Ar-C=O}$
e.g., methanoic acid (formic acid)	HCOOH
ethanoic acid (acetic acid)	CH_3COOH
octadecanoic acid (stearic acid)	$CH_3(CH_2)_{16}COOH$
Esters[b]	RCOOR' or $\underbrace{R-C}_{acid}\overset{O}{\overset{\|}{}}\underbrace{-O-R'}_{alcohol}$
e.g., methyl propanoate	$CH_3CH_2COOCH_3$

[a] The letter R represents an alkyl group and Ar represents an aromatic group
[b] The letter R represents hydrogen, alkyl group or aromatic group and R' represents alkyl group or aromatic group

arranged in a six-membered ring with one hydrogen atom bonded to each carbon atom). Some examples are shown in Table 39.2a.

Alcohols, Ethers, Phenols, and Thiols

Organic molecules with certain functional groups are synthesized for desirable properties. Alcohols, ethers, and phenols are derived from the structure of water by replacing the hydrogen atoms of water with alkyl groups (R) or aromatic (Ar) rings. For example, phenol is a class of compounds that has a hydroxy group attached to an aromatic ring (benzene ring). Organic compounds that contain the $-SH$ group are analogs of alcohols, and are known as thiols. Some examples are shown in Table 39.2b.

Aldehydes and Ketones

Both aldehydes and ketones contain the carbonyl group

$$\diagdown C = O,$$

a carbon–oxygen double bond. Aldehydes have at least one hydrogen atom bonded to the carbonyl group, whereas ketones have only alkyl or aromatic group bonded to the carbonyl group. The general formula for the saturated homologous series of aldehydes and ketones is $C_nH_{2n}O$. Some examples are shown in Table 39.2c.

Carboxyl Acids and Esters

The functional group of the carboxylic acids is known as a carboxyl group, represented as $-COOH$. Carboxylic acids can be either aliphatic (RCOOH) or aromatic (ArCOOH). The carboxylic acids with even numbers of carbon atoms n ranging from 4 to about 20 are called fatty acids (e.g., $n = 10, 12, 14, 16$, and 18, called capric acid, lauric acid, myristic acid, palmitic acid, and stearic acid, respectively).

Esters are alcohol derivates of carboxylic acids. Their general formula is RCOOR', where R may be

Table 39.2 (e) Names and formulas of selected organic nitrogen compounds (amides and amines)

Name	Formula
Amides	$RCONH_2$ or $\underset{\underset{NH_2}{\|}}{R-C=O}$
e.g., methanamide (formamide)	$HCONH_2$
ethanamide (acetamide)	CH_3CONH_2
Amines	RNH_2 or $R-N\overset{H}{\underset{H}{\diagdown}}$
	R_2NH
	R_3N
e.g., methylamine	CH_3NH_2
ethylamine	$CH_3CH_2NH_2$

The letter R represents an alkyl group or aromatic group

Table 39.3 (a) Some examples of polar (hydrophilic) and nonpolar (hydrophobic) groups

Name	Formula
Polar	
Alcohol (hydroxyl)	$-$OH
Carboxyl acid	$-$COOH
Aldehyde	$-$COH
Ketone	$R-\underset{\underset{O}{\|}}{C}-R$
Ester	$-$COO$-$
Carbonyl	$\diagup^{C=O}$
Ether	R$-$O$-$R
Amine	$-$NH$_2$
Amide	$-\underset{\underset{O}{\|}}{C}-NH_2$
Phenol	(benzene ring)$-$OH
Thiol	$-$SH
Trichlorosilane	SiCl$_3$
Nonpolar	
Methyl	$-$CH$_3$
Trifluoromethyl	$-$CF$_3$
Aryl (benzene ring)	(benzene ring)

The letter R represents an alkyl group

Table 39.3 (b) Organic groups listed in increasing order of polarity

Alkanes
Alkenes
Aromatic hydrocarbons
Ethers
Trichlorosilanes
Aldehydes, ketones, esters, carbonyls
Thiols
Amines
Alcohols, phenols
Amides
Carboxylic acids

a hydrogen, alkyl group or aromatic group, and R′ may be an alkyl group or aromatic group, but not a hydrogen. Esters are found in fats and oils. Some examples are shown in Table 39.2d.

Amides and Amines

Amides and amines are organic compounds containing nitrogen. Amides are nitrogen derivates of carboxylic acids. The carbon atom of a carbonyl group is bonded directly to a nitrogen atom of an $-$NH$_2$, $-$NHR or $-$NR$_2$ group. The characteristic structure of amide is RCONH$_2$.

An amine is a substituted ammonia molecule which has a general structure of RNH$_2$, R$_2$NH or R$_3$N, where R is an alkyl or aromatic group. Some examples are shown in Table 39.2e.

39.2.3 Polar and Nonpolar Groups

Table 39.3a summarizes polar and nonpolar groups commonly used in the construction of hydrophobic and hydrophilic molecules. Table 39.3b lists the relative polarity of selected polar groups [39.75]. Thiol, silane, carboxylic acid, and alcohol (hydroxyl) groups are the most commonly used polar anchor groups for their attachment to surfaces. Silane anchor groups are commonly used for Si or SiO$_2$ surfaces, as $-$Si$-$O$-$ bonds are strong. Methyl and trifluoromethyl are commonly used as end groups for hydrophobic film surfaces.

39.3 Self-Assembled Monolayers: Substrates, Spacer Chains, and End Groups in the Molecular Chains

SAMs are formed as a result of spontaneous, self-organization of functionalized organic molecules onto the surfaces of appropriate substrates into stable, well-defined structures (Fig. 39.3). The final structure is close to or at thermodynamic equilibrium, and as a result it tends to form spontaneously and rejects defects. SAMs consist of three building groups: a head group that binds strongly to a substrate, a surface terminal (tail or end) group that constitutes the outer surface of the film, and a spacer chain (backbone chain) that connects the head

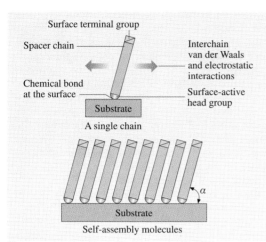

Fig. 39.3 Schematic of a SAM on a surface and the associated forces

and surface terminal groups. SAMs are named based on the surface terminal group, followed by the spacer chain and the head group (or type of compound formed at the surface). In order to control hydrophobicity, adhesion, friction, and wear, it should be strongly adherent to the substrate, and the surface terminal group of the organic molecular chain should be nonpolar. For strong attachment of the organic molecules to the substrate, the head group of the molecular chain should contain a polar terminal group, resulting in the exothermic process (energies on the order of tens of kcal/mol), i. e., the apparent pinning of the head group to a specific site on the surface through a chemical bond. Furthermore, the molecular structure and any cross-linking would have a significant effect on their friction and wear performance. The substrate surface should have a high surface energy (hydrophilic), so that there will be a strong tendency for molecules to adsorb onto the surface. The surface should be highly functional with polar groups and dangling bonds (generally unpaired electrons) so that they can react with organic molecules and provide a strong bond. Because of the exothermic head group–substrate interactions, molecules try to occupy every available binding site on the surface, and during this process they generally push together molecules that have already adsorbed. The process results in the formation of ordered molecular assemblies. The interactions between molecular chains are of van der Waals or electrostatic type, with energies on the order of a few kcal/mol (< 10), exothermic. The molecular chains in SAMs are not perpendicular to the surface; the tilt angle depends on the anchor group as well as on the substrate and the spacer group. For example, the tilt angle for alkanethiolate on Au is typically about $30-35°$ with respect to the substrate normal.

Table 39.4 lists selected systems which have been used for the formation of SAMs [39.44]. The spacer chain of the SAM is mostly an alkyl chain (($-CH_2)_n$) or made of a derivatized alkyl group. By attaching different terminal groups at the surface, the film surface can be made to attract or repel water. The commonly used surface terminal group of a hydrophobic film with low surface energy, in the case of a single alkyl chain, is a nonpolar methyl ($-CH_3$) or trifluoromethyl ($-CF_3$) group. For a hydrophilic film, the commonly used surface terminal groups are alcohol ($-OH$) or carboxylic acid ($-COOH$) groups. The surface active head groups most commonly used are thiol ($-SH$), silane (e.g., trichlorosilane or $-SiCl_3$), and carboxyl ($-COOH$) groups. The substrates most commonly used are gold, silver, platinum, copper, hydroxylated (activated) surfaces of SiO_2 on Si, Al_2O_3 on Al, and glass.

As an example substrate, epitaxial Au film on glass, mica or single-crystal silicon, produced by e-beam

Table 39.4 Selected substrates and precursors which have commonly been used for formation of SAMs

Substrate	Precursor	Binding with substrate
Au	RSH (thiol)	RS−Au
Au	ArSH (thiol)	ArS−Au
Au	RSSR′ (disulfide)	RS−Au
Au	RSR′ (sulfide)	−
Si/SiO$_2$, glass	RSiCl$_3$ (trichlorosilane)	Si−O−Si (siloxane)
Si/Si−H	RCOOH (carboxyl)	R−Si
Metal oxides (e.g., Al$_2$O$_3$, SnO$_2$, TiO$_2$)	RCOOH (carboxyl)	RCOO−···−MO$_n$

R represents alkane (C_nH_{2n+2}) and Ar represents aromatic hydrocarbon. It consists of various surface active headgroups and mostly with methyl terminal group

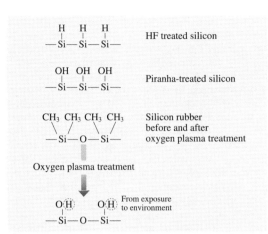

Fig. 39.4 Schematic showing HF-treated silica and the hydroxylation process occurring on a silica and elastomeric surfaces using Piranha solution and oxygen plasma, respectively

evaporation, is commonly used because it can be deposited on smooth surfaces as a film which is atomically flat and defect free. Bulk single-crystal Si, sputtered Al film, bulk Al sheets, and copper disks with natural oxide layers have been selected because of their use in the construction of MEMS/NEMS (e.g., digital projection displays) and heat-exchange applications, respectively. The substrate surface should be clean before deposition. For silicon substrates, a concentrated HF solution (typically 49% HF) is commonly used to remove the oxide layer, followed by a rinse with deionized water [39.60, 62]. Hydrogen passivates the surface by saturating the dangling bonds, which results in a hydrogen-terminated silicon surface with hydrophobic properties. For deposition of multimolecularly thick polymer films with nonpolar ends, hydrophobic substrates may lead to a coated surface with a high contact angle, which is preferred. For SAM deposition, the substrate should be hydrophilic in order to form strong interfacial bonds with their head groups. Hydroxylation of oxide surfaces is carried out to make them hydrophilic. Silicon and other metals get oxidized and get hydroxylated to some degree when exposed to the environment. Bulk silicon, polysilicon film or SiO_2 film surfaces are commonly treated to produce a hydroxylated silica surface by immersion in Piranha solution (a mixture of typically 3 : 1 v/v 98%H_2SO_4 : 30%H_2O_2) at temperatures of $\approx 90\,°C$ for $\approx 30\,min$ followed by a rinse in deionized (DI) water [39.27, 60, 62]. Piranha solution also removes any organic and metallic contaminants, whereas HF would not necessarily remove organics. Oxygen plasma is another technique used for hydroxylation for SiO_2 as well as polymer surfaces [39.59, 61, 64, 73]. For complex silicon geometries or fine structures, such as AFM tips, oxygen plasma may be preferable. Figure 39.4 shows the schematics of surfaces after various surface treatments. Surfaces after piranha or oxygen plasma treatment remain hydrophilic for a few hours to about a day and become hydrophobic when they come into contact with carbon. They can retain hydrophilicity longer in dry nitrogen. To retain hydrophobicity, polymers are generally stored in DI water. Surfaces treated with HF remain hydrophobic for $\approx 2-3\,h$ and can retain hydrophobicity longer in dry nitrogen.

For organic molecules to pack together and provide a better ordering, a substrate for given molecules should be selected such that the cross-sectional diameter of the spacer chains of the molecule is equal to or smaller than the distance between the anchor groups attached to the substrate. For the case of alkanethiol film, the advan-

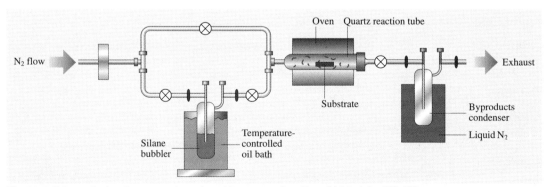

Fig. 39.5 Schematic showing vapor-phase deposition system for silane SAMs (after [39.59])

tage of Au substrate over SiO_2 substrate is that it results in better ordering because the cross-sectional diameter of the alkane molecule is slightly smaller than the distance between sulfur atoms attached to the Au substrate (≈ 0.53 nm). The thickness of the film can be controlled by varying the length of the hydrocarbon chain, and the surface properties of the film can be modified by the terminal group.

SAMs are usually produced by immersing a substrate in a solution containing the precursor (ligand) that is reactive to the substrate surface or by exposing the substrate to the vapor of the reactive chemical precursors [39.28]. A schematic of the vapor deposition system is shown in Fig. 39.5 [39.59, 60, 73]. Samples are placed in the quartz reaction tube. The silane bubbler is used for introducing gas-phase silane into the quartz reaction tube placed in an oven at a controlled temperature. An inert gas flow (N_2) is used as a carrier gas. A byproducts condenser is used for trapping the byproducts and/or nonreacted silanes.

Research on some SAMs has been widely reported. SAMs of long-chain fatty acids $C_nH_{2n+1}COOH$ or $(CH_3)(CH_2)_nCOOH$ ($n = 10, 12, 14$ or 16) on glass or alumina substrates have been widely studied since the 1950s [39.24, 25, 28, 29, 76]. Probably the most studied SAMs to date are n-alkanethiolate (n-alkyl and n-alkane are used interchangeably) monolayers $CH_3(CH_2)_nS-$ prepared from the adsorption of alkanethiol $-(CH_2)_nSH$ solution onto a Au film [39.44, 77–81] and n-alkylsiloxane. *Siloxane* ($Si-O-Si$) refers to the bond, whereas *silane* (Si_nX_{2n+2}, which includes a covalently bonded compound containing the ele-

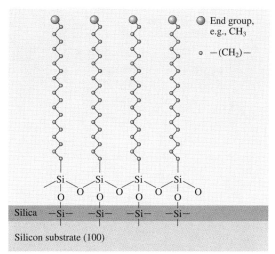

Fig. 39.6 Schematics of a methyl-terminated, n-alkylsiloxane monolayer on Si/SiO_2

ments Si and other atoms or groups such as H and Cl to form SiH_4 and $SiCl_4$, respectively) refers to the head group of the precursor. These terms are used interchangeably. Monolayers produced by adsorption of n-alkyltrichlorosilane $-(CH_2)_nSiCl_3$ onto a hydroxylated Si/SiO_2 substrate with siloxane ($Si-O-Si$) binding (Fig. 39.6) [39.52, 58–63, 82]. *Tambe* and *Bhushan* [39.52], *Bhushan* et al. [39.53], *Hoque* et al. [39.54–56, 65–67]. *DeRose* et al. [39.57] have produced perfluoroalkylsilane and perfluoroalkylphosphonate on Al and Cu surfaces.

39.4 Contact Angle and Nanotribological Properties of SAMs

The basis for the molecular design and tailoring of SAMs should start from complete knowledge of the interrelationships between the molecular structure and contact angle and nanotribological properties of SAMs, as well as deep understanding of the adhesion, friction, and wear mechanisms of SAMs at the molecular level. Friction and wear studies of SAMs have been carried out on macro- and nanoscales. Macroscale tests are conducted using a so-called pin-on-disk tribotester apparatus in which a ball specimen slides against a lubricated flat specimen [39.14, 15]. Nanoscale tests are conducted using an atomic force/friction force microscope (AFM/FFM) [39.4, 6, 14, 15]. In AFM/FFM experiments, a sharp tip of a radius ranging from ≈ 5 to 50 nm slides against a SAM specimen. A Si_3N_4 tip is commonly used for friction studies, and a Si or natural diamond tip is commonly used for scratch, wear, and indentation studies.

In early studies, the effect of chain length of the carbon atoms of fatty-acid monolayers on the coefficient of friction and wear on the macroscale was studied by *Bowden* and *Tabor* [39.24] and *Zisman* [39.25]. *Zisman* [39.25] reported that, for monolayers deposited on a glass surface sliding against a stainless-steel surface, there is a steady decrease in friction with increasing chain length. At a significantly long chain length, the coefficient of friction reaches a lower limit (Fig. 39.7a). He further reported that monolayers having a chain

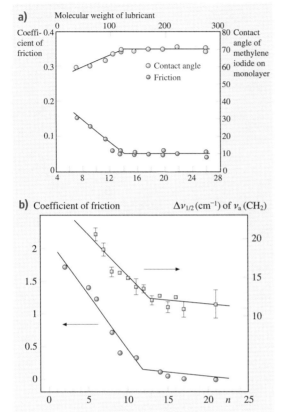

Fig. 39.7 (a) Effect of chain length (or molecular weight) on the coefficient of macroscale friction of stainless steel sliding on glass lubricated with a monolayer of fatty acid, and contact angle of methyl iodide on condensed monolayers of fatty acid on glass (after [39.25]). (b) Effect of chain length of methyl-terminated n-alkanethiolate over Au film AuS(CH$_2$)$_n$CH$_3$ on the coefficient of microscale friction and peak bandwidth at half maximum ($\Delta \nu_{1/2}$) for the bandwidth of the methylene stretching mode (ν_a(CH$_2$)) (after [39.85])

length below 12 C atoms behave as liquids (poor durability), and those with a chain length of 12–15 C atoms behave like a plastic solid (medium durability), whereas those with a chain length above 15 C atoms behave like a crystalline solid (high durability). Investigations by *Ruhe* et al. [39.83] indicated that the lifetime of a alkylsilane monolayer coating on a silicon surface increases greatly with increasing chain length of the alkyl substituent. *DePalma* and *Tillman* [39.84] showed that a monolayer of n-octadecyltrichlorosilane (n-C$_{18}$H$_{37}$SiCl$_3$, OTS) is an effective lubricant on silicon.

With the development of AFM techniques, researchers have successfully characterized the nanotribological properties of self-assembled monolayers [39.1, 4, 6]. Studies by *Bhushan* et al. [39.58] showed that C$_{18}$ alkylsiloxane films exhibit the lowest coefficient of friction and can withstand a much higher normal load during sliding as compared with LB films, soft Au films, and hard SiO$_2$ coatings. *McDermott* et al. [39.85] used AFM to study the effect of the length of alkyl chains on the frictional properties of methyl-terminated n-alkylthiolate CH$_3$(CH$_2$)$_n$S– films chemisorbed on Au(111). They reported that longer-chain monolayers exhibit markedly lower friction and reduced propensity for wear than shorter-chain monolayers (Fig. 39.7b). These results are in good agreement with the macroscale results by *Zisman* [39.25]. They also conducted infrared reflection spectroscopy to measure the bandwidth of the methylene stretching mode (ν_a(CH$_2$)) which exhibits a qualitative correlation with the packing density of the chains. It was found that the chain structures of monolayers prepared with longer chain lengths are more ordered and more densely packed in comparison with those of monolayers prepared with shorter chain lengths. They further reported that the ability of the longer-chain monolayers to retain molecular-scale order during shear leads to a lower observed friction. Monolayers having a chain length of more than 12 C atoms, preferably 18 or more, are desirable for tribological applications. (Incidentally, monolayers with 18 C atoms, octadecanethiol films, have been widely studied.)

Xiao et al. [39.86] and *Lio* et al. [39.87] also studied the effect of the length of the alkyl chains on the frictional properties of n-alkanethiolate films on gold and n-alkylsilane films on mica. Friction was found to be particularly high with short chains of fewer than eight carbon atoms. Thiols and silanes exhibit similar friction forces for the same n when $n > 11$, while for $n < 11$, silanes exhibit higher friction, larger than that for thiols by a factor of about 3 for $n = 6$. The increase in friction was attributed to the large number of dissipative modes in the less ordered chains that occurs when going from a thiol to a silane anchor or when decreasing n. Longer chains ($n > 11$), stabilized by van der Waals attraction, form more compact and rigid layers and act as better lubricants. *Schönherr* and *Vancso* [39.88] also correlated the magnitude of friction with the order of the alkane chains. The disorder of short-chain hydrocarbon disulfide SAMs was found to result in a significant increase in the magnitude of friction.

Tsukruk and *Bliznyuk* [39.89] studied the adhesion and friction between a Si sample and a Si_3N_4 tip, in which both surfaces were modified by $-CH_3$-, $-NH_2$-, and $-SO_3H$-terminated silane-based SAMs. Various polymer molecules were used for the backbone. They reported a very broad maximum adhesive force in the pH range from 4 to 8, with minimum adhesion at pH > 9 and pH < 3 for all of the studied mating surfaces. This observation can be understood by considering a balance of electrostatic and van der Waals interactions between composite surfaces with multiple isoelectric points. The friction coefficient of NH_2/NH_2 and SO_3H/SO_3H mating SAMs was very high in aqueous solutions. Cappings of NH_2-modified surfaces (3-aminopropyltriethoxysilane) with rigid and soft polymer layers resulted in a significant reduction in adhesion to a level lower than that of untreated surface [39.90]. *Fujihira* et al. [39.91] studied the influence of surface terminal groups of SAMs and functional tip on adhesive force. It was found that the adhesive forces measured in air increase in the order: CH_3/CH_3, $CH_3/COOH$, $COOH/COOH$.

Bhushan and *Liu* [39.79], *Liu* et al. [39.80], and *Liu* and *Bhushan* [39.81, 92] studied adhesion, friction, and wear properties of alkylthiol and biphenylthiol SAMs on Au(111) films. They explained the friction mechanisms using a molecular spring model in which local stiffness and intermolecular forces govern the friction properties. They studied the influence of relative humidity, temperature, and velocity on adhesion and friction. They also investigated the wear mechanisms of SAMs by a continuous microscratch AFM technique.

Fluorinated carbon (fluorocarbon) molecules are known to have low surface energy and are commonly used for lubrication [39.14, 15]. *Bhushan* and *Cichomski* [39.73] deposited fluorosilane SAMs on polydimethylsiloxane (PDMS). To make a hydrophobic PDMS surface chemically active, PDMS surface was oxygenated using an oxygen plasma, which introduces silanol groups (SiOH). They reported that SAM-coated PDMS was more hydrophobic, with lower adhesion, friction, and wear. *Bhushan* et al. [39.59, 60], *Kasai* et al. [39.61], *Lee* et al. [39.62], *Tambe* and *Bhushan* [39.52], and *Tao* and *Bhushan* [39.64] studied the adhesion, friction, and wear of methyl- and/or perfluoro-terminated alkylsilanes on silicon. They reported that perfluoroalkylsilane SAMs exhibited lower surface energy, higher contact angle, lower adhesive force, and lower wear as compared with alkylsilanes.

Kasai et al. [39.61] also reported the influence of relative humidity, temperature, and velocity on adhesion and friction. *Tao* and *Bhushan* [39.63] studied degradation mechanisms of alkylsilanes and perfluoroalkylsilane SAMs on Si. They reported that oxygen in the air causes thermal oxidation of SAMs.

Tambe and *Bhushan* [39.52], *Bhushan* et al. [39.53], *Hoque* et al. [39.54, 55], and *DeRose* et al. [39.57] studied the nanotribological properties of methyl- and perfluoro-terminated alkylphosphonate, perfluorodecyldimethylchlorosilane, and perfluorodecanoic acid on aluminum, of industrial interest. *Hoque* et al. [39.56] and *DeRose* et al. [39.57] studied the nanotribological properties of alkylsilanes and perfluroalkylsilanes on aluminum. *Hoque* et al. [39.65–67] studied the nanotribological properties of alkylphosphonate and perfluoroalkylsilane SAMs on copper. The authors found that these SAMs on aluminum and copper perform well irrespective of the substrate used. They confirmed the presence of respective films using x-ray photoelectron spectroscopy (XPS).

Hoque et al. [39.65–67] studied the chemical stability of various SAMs deposited on Cu substrates via exposure to various corrosive conditions. *DeRose* et al. [39.57] studied the chemical stability of various SAMs deposited on Al substrates via exposure to corrosive conditions (aqueous nitric acid solutions of low pH of 1.8 at temperatures ranging from 60 °C to 80 °C for times ranging from 30 to 70 min). The exposed samples were characterized by XPS and contact angle measurements. They reported that perfluorodecanoic acid/Al is less stable than perfluorodecylphosphonate/Al and octadecylphosphonate/Al, but more stable than perfluorodecyldimethylchlorosilane/Al, which has implications in DMD applications, discussed earlier. In general, chemical stability data of various SAMs deposited on Cu and Al surfaces to corrosive environments has been reported by these authors. Based on these studies, it was concluded that chemisorption occurs at the interface and is responsible for strong interfacial bonds.

To date, the contact angle and nanotribological properties of alkanethiol, biphenylthiol, alkylsilane, perfluoroalkylsilane, alkylphosphonate, and perfluoroalkylphosphane SAMs have been widely studied. In this chapter, we review in some detail the nanotribological properties of various SAMs having alkyl and biphenyl spacer chains with different surface terminal groups ($-CH_3$ and $-CF_3$) and head groups ($-S-H$, $-Si-O-$, $-OH$, and $P-O-$) which have been investigated by AFM at various operating conditions

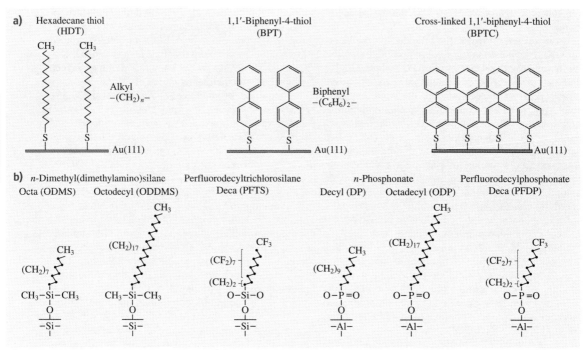

Fig. 39.8a,b Schematics of the structures of (**a**) hexadecane and biphenyl thiol SAMs on Au(111) substrates, and (**b**) perfluoroalkylsilane and alkylsilane SAMs on Si with native oxide substrates, and perfluoroalkylphosphonate and alkylphosphonate SAMs on Al with native oxide

(Fig. 39.8a,b) [39.52, 53, 59, 61, 63, 79–81, 92]. Hexadecane thiol (HDT), 1,1′-biphenyl-4-thiol (BPT), and cross-linked BPT (BPTC) were deposited on Au(111) films on Si(111) substrates by immersing the substrate in a solution containing the precursor (ligand) that is reactive to the substrate surface. Cross-linked BPTC was produced by irradiation of BPT monolayers with low-energy electrons. Perfluoroalkylsilane and alkylsilane SAMs were deposited on Si(100) by exposing the substrate to the vapor of the reactive chemical precursors. Perfluoroalkylphosphonate and alkylphosphonate SAMs were deposited on sputtered Al film on Si substrate as well as bulk Al substrates. Representative data follow.

39.4.1 Measurement Techniques

Experimental techniques used for measurement of the static contact angle, surface energy, adhesion, friction, and wear are described next.

Static Contact Angle and Surface Energy Measurements Using DI Water

The static contact angle, a measure of water-repellent property, was measured using a Rame–Hart model 100 contact angle goniometer (Mountain Lakes, NJ) [39.93, 94]. Typically, $10\,\mu l$ droplets of DI water were used for making contact angle measurements. At least two measurements of the contact angle were made and were found to be reproducible within $\pm 2°$. The critical surface tension, a measure of interfacial surface energy, was obtained from the so-called Zisman plot. Contact angles of SAMs with liquids with a range of surface tensions were measured. The cosines of the contact angles were plotted as a function of surface tension of the various n-alkane liquids (hexadecane, dodecane, undecane, and decane) used. The plot is linear for low-polarizable SAMs. The horizontal intercept of the line passing through cos (contact angle) = 1 provides the critical surface tension, which is a measure of the surface energy of the SAM [39.59].

AFM Adhesion and Friction Measurements

The adhesion and friction tests were conducted using a commercial AFM system. Square-pyramidal Si_3N_4 tips with a 30–50 nm tip radius on gold back-coated triangular Si_3N_4 cantilevers with a typical spring constant of 0.58 N/m were used. Adhesion can be calculated using either force calibration plots or from the negative intercepts on plots of friction force versus normal loads. Both methods generally yield similar results. The force calibration plot technique was used in this study. The coefficient of friction was obtained from the slope of plots of friction force versus normal load. Normal loads typically ranged from 5 to 100 nN. Friction force measurements were generally performed at a scan rate of 1 Hz along the fast scan axis and over a scan size of $2 \times 2\,\mu m^2$. The fast scan axis was perpendicular to the longitudinal direction of the cantilever. The friction force was calibrated by the method described in [39.4, 6].

AFM Wear Measurements

Wear tests were conducted using a diamond tip with a nominal radius of 50 nm and nominal cantilever stiffness of 10 N/m. Wear tests were performed on a $1 \times 1\,\mu m^2$ scan area at the desired normal load and at a scan rate of 1 Hz. After each wear test, a $3 \times 3\,\mu m^2$ area was imaged, and the average wear depth was calculated.

Effect of Relative Humidity, Temperature, and Sliding Velocity

The influence of relative humidity on adhesive force, friction force, and wear was studied in an environmentally controlled chamber. Relative humidity was controlled by introducing a mixture of dry and moist air inside the chamber. The temperature was maintained at $22 \pm 1\,°C$. The sample was kept in the environmental chamber at desired humidity for at least 2 h prior to the tests so that the system could reach equilibrium condition.

In order to study the effect of temperature on adhesion and friction force, the samples were placed on a thermal stage during the measurements. A glass plate was placed under the thermal stage to prevent heat from being transported away. The temperature range studied was from $20\,°C$ to $110\,°C$. The relative humidity was maintained at $50 \pm 5\%$ during the temperature effect measurements.

The effect of sliding velocity on friction force was monitored in ambient conditions using a high-velocity piezoelectric stage designed for achieving high relative sliding velocities on a commercial AFM setup [39.95]. The traveling distance of the sample, i. e., the scan size, was set at $25\,\mu m$, while the scan frequency was varied between 0.1 Hz ($5\,\mu m/s$) and 100 Hz ($5000\,\mu m/s$).

Chemical Degradation and Environmental Studies

The chemical degradation experiments were carried out in a high-vacuum tribotest apparatus [39.96, 97]. The system was equipped with a mass spectrometer so that gaseous emissions from the interface could be monitored in situ during the sliding in high vacuum and other controlled environments. The normal loads and friction forces at the contacting interface were measured using resistive-type strain-gage transducers. For the sliding tests, the coated flat sample was glued onto a flat surface at the end of a rotating shaft. The sample was slid against a Si(100) wafer mounted on the flat surface of a slider integrated with a flexible cantilever used in magnetic rigid disk drives. The sliding speed used was 0.3 m/s, and the applied pressure was 50 kPa. The environmental effects were investigated in high vacuum (2×10^{-7} Torr), argon, dry air (less than 2% RH), ambient air (30% RH), and high-humidity air (70% RH).

39.4.2 Hexadecane Thiol and Biphenyl Thiol SAMs on Au(111)

Hexadecane thiol on Au(111) film was selected as it is a widely studied film. Biphenyl thiol was selected to study the effect of rigidity on nanotribological performance. Biphenyl thiol film was cross-linked to further increase its stiffness.

Surface Roughness, Adhesion, and Friction

Surface height and friction force images of SAMs were recorded simultaneously on an area of $1 \times 1\,\mu m^2$ by an AFM, and adhesive forces were measured by using the force calibration mode in an AFM [39.79].

For further analysis, presented later in this chapter, the measured roughness, thickness, tilt angles, and spacer chain lengths of Si(111), Au(111), and various SAMs are listed in Table 39.5 [39.79]. The roughness of BPT is very close to that of Au(111), but the roughness of BPTC is lower than that of Au(111) and BPT; this is caused by electron irradiation. Table 39.5 indicates that the roughness value of HDT is much higher than the substrate roughness for Au(111). This is caused by local aggregation of organic compounds on the substrates during SAMs deposition. Table 39.5 also indicates that the thickness of biphenyl thiol SAMs are generally

Table 39.5 The R_a roughness, thickness, tilt angles, and spacer chain lengths of SAMs

Samples	R_a roughness[a] (nm)	Thickness[b] (nm)	Tilt angle[b] (deg)	Spacer length[c] (nm)
Si(111)	0.07	–	–	–
Au(111)	0.37	–	–	–
HDT	0.92	1.89	30	1.91
BPT	0.36	1.25	15	0.89
BPTC	0.14	1.14	25	0.89

[a] Measured by an AFM with $1 \times 1\ \mu m^2$ scan size, using Si_3N_4 tip under 3.3 nN normal load
[b] The thickness and tilt angles of BPT and BPTC are reported by *Geyer* et al. [39.78]. The thickness and tilt angles of HDT are reported by *Ulman* [39.29].
[c] The spacer chain lengths of alkylthiols were calculated by the method reported by *Miura* et al. [39.98]. The spacer chain lengths of biphenyl thiols were calculated by the data reported by *Ratajczak-Sitarz* et al. [39.99]

smaller than the alkylthiol, because of the shorter spacer chains in biphenyl thiol.

The average values and standard deviation of the adhesive force and coefficient of friction are presented in Fig. 39.9 [39.79]. Based on the data, the adhesive force and coefficient of friction of SAMs are less than those of their corresponding substrates. Among various films, HDT exhibits the lowest values. The ranking of adhesive forces F_a is in the following order: $F_{a\text{-Au}} > F_{a\text{-BPT}} > F_{a\text{-BPTC}} > F_{a\text{-HDT}}$, and the ranking of the coefficients of friction μ is in the following order: $\mu_{Au} > \mu_{BPTC} > \mu_{BPT} > \mu_{HDT}$. The ranking of various SAMs for adhesive force and coefficient of friction are similar. This suggests that alkylthiol and biphenyl thiol SAMs can be used as effective molecular lubricants for micro-/nanodevices.

In micro-/nanoscale contact, liquid capillary condensation is one of the sources of adhesion and friction. In the case of a sphere in contact with a flat surface, the attractive Laplace force caused by water capillary forces is [39.14, 15]

$$F_L = 2\pi R \gamma_{la}(\cos\theta_1 + \cos\theta_2), \quad (39.1)$$

where R is the radius of the sphere, γ_{la} is the surface tension of the liquid against air, and θ_1 and θ_2 are the contact angles between liquid and flat and spherical surfaces, respectively. In an AFM adhesive study, the tip–flat sample contact is just like a sphere in contact with a flat surface, and the liquid is water. Since a single tip was used in the adhesion measurements, $\cos\theta_2$ can be treated as a constant. Therefore,

$$F_L = 2\pi R \gamma_{la}(1 + \cos\theta_1) - 2\pi R \gamma_{la}(1 - \cos\theta_2)$$
$$= 2\pi R \gamma_{la}(1 + \cos\theta_1) - C, \quad (39.2)$$

where C is a constant.

Based on the following Young–Dupré equation, the work of adhesion W_a (the work required to pull apart a unit area of solid–liquid interface) can be written as [39.12]

$$W_a = \gamma_{la}(1 + \cos\theta_1). \quad (39.3)$$

It indicates that W_a is determined by the contact angle of SAMs, i.e., is influenced by the surface chemistry properties (polarization and hydrophobicity) of SAMs. By substituting (39.3) into (39.2), F_L can be expressed as

$$F_L = 2\pi R W_a - C. \quad (39.4)$$

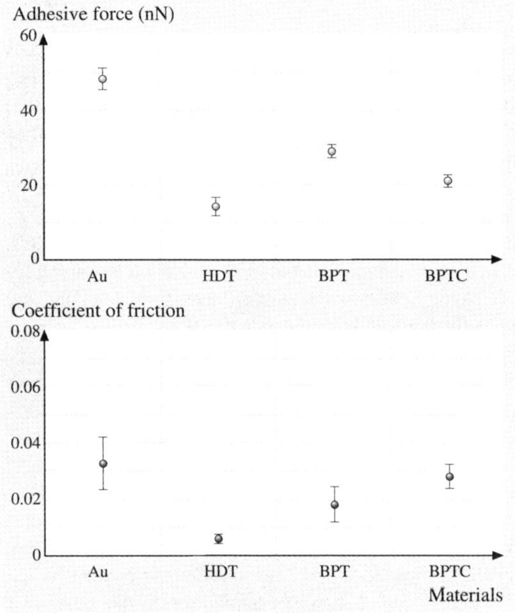

Fig. 39.9 Adhesive forces and coefficients of friction of Au(111) and various SAMs

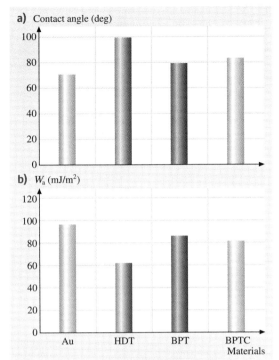

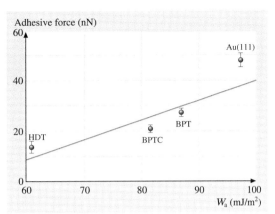

Fig. 39.11 Relationship between the adhesive force and work of adhesion of different specimens

Fig. 39.10 (a) The static contact angle, and **(b)** work of adhesion of Au(111) and various SAMs. All of the points in this figure represent the mean of six measurements. The uncertainty associated with the average contact angle is within ±2°

When the influence of other factors, such as the van der Waals force, on the adhesive force are very small, the adhesive force $F_a \approx F_L$. Thus the adhesive force F_a should be proportional to the work of adhesion W_a.

Contact angle is a measure of the wettability of a solid by a liquid and determines the W_a value [39.93, 94]. The static contact angles of distilled water on Au(111) and SAMs were measured and are summarized in Fig. 39.10a [39.79]. For water, $\gamma_{la} = 72.6 \text{ mJ/m}^2$ at 22 °C, and by using (39.3), the W_a data are obtained and presented in Fig. 39.10b. The W_a can be ranked in the following order: $W_{a\text{-Au}}$ (97.1) > $W_{a\text{-BPT}}$ (86.8) > $W_{a\text{-BPTC}}$ (82.1) > $W_{a\text{-HDT}}$ (61.4). Except $W_{a\text{-Au}}$, this order exactly matches the order of adhesion force in Fig. 39.9. The relationship between F_a and W_a is summarized in Fig. 39.11 [39.79]. This indicates that the adhesive force F_a (nN) increases with the work of adhesion W_a (mJ/m^2) according to the following linear relationship

$$F_a = 0.57 W_a - 22 . \tag{39.5}$$

These experimental results agree well with the modeling prediction presented earlier (39.4). This proves that, on the nanoscale at ambient conditions, the adhesive force of SAMs is mainly influenced by the water capillary force. Though neither HDT nor BPT has polar surface groups, the surface terminal of HDT has a symmetrical structure, which causes a smaller electrostatic attractive force and yields a smaller adhesive force than BPT. It is believed that the easy attachment of Au onto the tip should be one of the reasons that causes the large adhesive force, which does not fit the linear relationship described by (39.5).

Stiffness, Molecular Spring Model, and Micropatterned SAMs

Next the friction mechanisms of SAMs were examined. Monte Carlo simulation of the mechanical relaxation of $CH_3(CH_2)_{15}SH$ self-assembled monolayer performed by *Siepman* and *McDonald* [39.100] indicated that SAMs compress and respond nearly elastically to microindentation by an AFM tip when the load was below a critical normal load. Compression can lead to major changes in the mean molecular tilt (i.e., orientation), but the original structure is recovered as the normal load is removed. *Garcia-Parajo* et al. [39.101] also reported compression and relaxation of octadecyltrichlorosilane (OTS) film during loading and unloading.

To study the difference in the stiffness of various films, the stiffness properties were measured by an AFM in force modulation mode [39.4, 6, 81]. It was re-

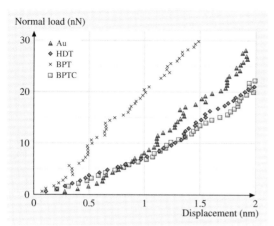

Fig. 39.12 Normal load versus displacement curves of Au(111) and various SAMs

ported that BPT was stiffer than HDT. Since BPT has a rigid benzene structure, it is more difficult to compress than HDT. Figure 39.12 shows the variation of the displacement with normal load in indentation mode, clearly indicating that SAMs can be compressed. At a given normal load, SAMs with long carbon chain structure such as HDT are easy to compress as compared with SAMs with rigid benzene ring structure such as BPT, which implies that BPT is more rigid than HDT.

In order to explain the frictional difference of SAMs, based on the friction and stiffness measurements by AFM and the Monte Carlo simulation, a molecular spring model is presented in Fig. 39.13. It is believed

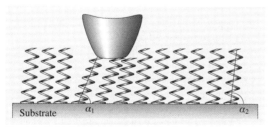

Fig. 39.13 Molecular spring model of a SAM. In this figure, $\alpha_1 < \alpha_2$, which is caused by the orientation under the normal load applied by the AFM tip. The orientation of the molecular springs reduces the shearing force at the interface, which in turn reduces the friction force. The molecular spring constant as well as the intermolecular forces can determine the magnitude of the coefficient of friction of the SAM. In this figure, the size of the tip and molecular springs are not to exact scale (after [39.79])

that the self-assembled molecules on a substrate are just like assembled molecular springs anchored to the substrates [39.79]. An AFM tip sliding on the surface of SAMs is like a tip sliding on the top of *molecular springs or brush*. The molecular spring assembly has compliant features and can experience compression and orientation under normal load. The orientation of the molecular springs or brush reduces the shearing force at the interface, which in turn reduces the friction force. The possibility of orientation is determined by the spring constant of a single molecule (local stiffness), as well as the interaction between neighboring molecules, which can be reflected by the packing density or packing energy. It should be noted that the orientation can lead to conformational defects along the molecular chains, which leads to energy dissipation. In the study of BPT by AFM, it was found that, after the first several scans, the friction force is significantly reduced, but the surface height does not show any apparent change. This suggests that molecular orientation can be facilitated by initial sliding and is reversible [39.92].

Based on the stiffness measurements obtained using Fig. 39.12 and the view of the molecular structure in Fig. 39.13, biphenyl is a more rigid structure due to the contribution of two rigid benzene rings. Therefore the spring constant of BPT is larger than that of HDT. The hydrogen (H^+) in a biphenyl chain has an electrostatic attractive force with the π electrons in the neighboring benzene ring. Thus the intermolecular force between biphenyl chains is stronger than that for alkyl chains. The larger spring constant of BPT and stronger intermolecular force require a larger external force to allow it to orient, thus causing a higher coefficient of friction. The cross-linking of BPT leads to a larger packing energy for BPTC. Therefore it requires a larger external force to allow BPTC orientation; i.e., the coefficient of BPTC is higher than BPT.

An elegant way to demonstrate the influence of molecular stiffness on friction is to investigate SAMs with different structures on the same wafer. For this purpose, a micropatterned SAM was prepared. First biphenyldimethylchlorosilane (BDCS) was deposited on silicon by a typical self-assembly method [39.81]. Then the film was partially cross-linked using a mask technique by low-energy electron irradiation. Finally the micropatterned BDCS films were realized, which had the as-deposited and cross-linked coating regions on the same wafer. The local stiffness properties of these micropatterned samples were investigated by force-modulation AFM technique [39.102]. The variation in the deflection amplitude provides a measure

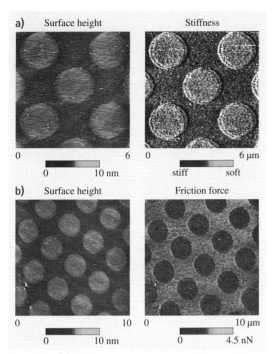

Fig. 39.14 (a) AFM grayscale surface height and stiffness images, and (b) AFM grayscale surface height and friction force images of micropatterned BDCS (after [39.81])

of the relative local stiffness of the surface. Surface height, stiffness, and friction images of the micropatterned biphenyldimethylchlorosilane (BDCS) specimen were obtained and are presented in Fig. 39.14 [39.81]. The circular areas correspond to the as-deposited film and the remaining area to the cross-linked film. Figure 39.14a indicates that cross-linking caused by the low-energy electron irradiation leads to about a 0.5 nm decrease of the surface height of BDCS films. The corresponding stiffness images indicate that the cross-linked area has a higher stiffness than the as-deposited area. Figure 39.14b indicates that the as-deposited area (area with higher surface height) has a lower friction force. Obviously, these data from the micropatterned sample prove that the local stiffness of SAMs has an influence on their friction performance; higher stiffness leads to larger friction force. These results provide strong proof of the suggested molecular spring model.

In summary, it is found that SAMs exhibit compliance and can experience compression and orientation under normal load. The orientation of SAMs reduces the shear stress at the interface; therefore SAMs can serve as good lubricants. The molecular spring constant (local stiffness), as well as the intermolecular forces, can influence the magnitude of the coefficients of friction of SAMs.

Wear and Scratch Resistance

Wear resistance was studied on an area of $1 \times 1\,\mu m^2$. The variation of wear depth with normal loads is presented in Fig. 39.15 [39.79]. HDT exhibits the best wear resistance. For all of the tested SAMs, in the curves of wear depth as a function of normal load, there appears a critical normal load, marked by arrows in Fig. 39.15. When the normal load is smaller than this critical normal load, the monolayer shows only a slight height change in the scan areas. When the normal load is higher than this critical value, the height change of the SAM increases dramatically. Relocation and accumulation of BPT molecules have been observed during the initial several scans, which lead to the formation of a larger terrace. Wear studies of a single BPT terrace indicate that the wear life of BPT increases exponentially with terrace size [39.80, 81].

Scratch resistance of Au(111) and SAMs were studied by a continuous AFM microscratch technique. Figure 39.16a shows coefficient of friction profiles as a function of increasing normal load, and corresponding tapping-mode AFM surface height images of the scratches captured on Au(111) and SAMs [39.81]. Figure 39.16a indicates that there is an abrupt increase in the coefficient of friction for all of the tested samples. The normal load associated with this event is termed the critical load (indicated by the arrows labeled A). At the initial stages of the scratch, all the samples exhibit a low coefficient of friction, indicating that the

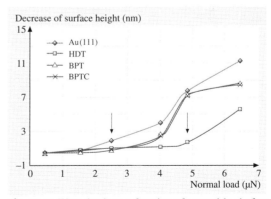

Fig. 39.15 Wear depth as a function of normal load after one scan cycle

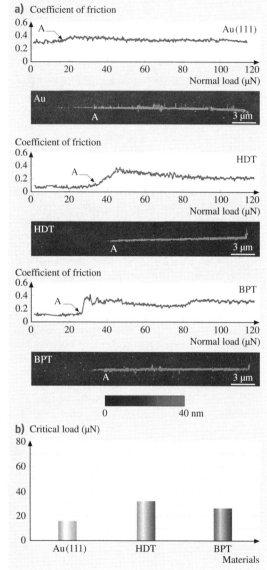

Fig. 39.16 (a) Coefficient of friction profiles during scratch as a function of normal load and corresponding AFM surface height images. (b) Critical loads estimated from the coefficient of friction profile and AFM images for Au(111), HDT/Au(111), and BPT/Au(111) (after [39.81])

friction force is dominated by the shear component. This is in agreement with analysis of the AFM images, which shows negligible damage on the surfaces prior to the critical load. At the critical load, a clear groove is formed, which is accompanied by the formation of material pileup at the sides of the scratch. This suggests that the initial damage that occurs at the critical load is due to plowing associated with plastic deformation, and this causes the sharp rise in the coefficient of friction. Beyond the critical load, debris can be seen in addition to material pileup at the sides of the scratch. Figure 39.16b summarizes the critical loads for the various samples obtained in this study. It clearly indicates that all SAMs can increase the critical load of the corresponding substrate.

The mechanisms responsible for a sudden drop in decrease in surface height with an increase in load during wear and scratch test need to be understood. *Barrena* et al. [39.103] observed that the height of self-assembled alkylsilanes decreases in discrete amounts with normal load. This step-like behavior is due to the discrete molecular tilts, which are dictated by the geometrical requirements of the close packing of molecules. Only certain angles are allowed due to the zigzag arrangement of the C atoms. The relative height of the monolayer under pressure can be calculated by the following equation

$$\left(\frac{h}{L}\right) = \left[1 + \left(\frac{na}{d}\right)^2\right]^{-1/2} , \qquad (39.6)$$

where L is the total length of the molecule, h is the height of the SAMs in the tilt configuration (monolayer thickness), a is the distance between alternate C atoms in the molecule, d is the separation of the molecules, and n is the step number. Values of a of 0.25 nm and d of 0.47 nm are used in the calculation for HDT. The calculated and measured relative heights of HDT are listed in Table 39.6. When the normal loads are smaller than the critical values in Fig. 39.15, the measured relative height values of HDT are very close to the calculated values. This means that HDT underwent step tilting below critical normal loads.

The residual SAM thickness after wear under critical normal load was measured by profiling the worn film using AFM. The results are listed in Table 39.7. For an alkanethiol monolayer, the relationship between the monolayer thickness h and intercept length L_0 can be expressed as (Fig. 39.17)

$$h = b\cos(\alpha)n + L_0 , \qquad (39.7)$$

where b is the length of the projection of the C–C bond onto the main chain axis ($b = 0.127$ nm for alkanethiol), n is the chain length defined by $CH_3(CH_2)_n SH$, and

Table 39.6 Calculated $[1+(na/d)^2]^{-1/2}$ and measured (h/L) relative heights of HDT self-assembled monolayer (after [39.79])

Steps (n)	Calculated[a] $[1+(na/d)^2]^{-1/2}$	Measured (h/L)
1	0.883	–
2	0.685	0.674[b]
3	0.531	0.532[c]
4	0.425	0.416[d]
5	0.352	0.354[e]
6	0.299	–

[a] Calculations are based on the assumption that the molecules tilt in discrete steps (n) upon compression with a diamond AFM tip [39.103]
[b–e] These measured values correspond to the normal loads of 0.50, 1.57, 2.53, and 4.03 µN, respectively

Table 39.7 Calculated L_0 and measured residual film thickness for SAMs under critical load

	L_0^a (nm)	Residual thickness[b] (nm)
HDT	0.24	0.25
BPT	0.39	0.42
BPTC	0.33	0.38

[a] Calculated by the equation: $h = b\cos(\alpha)n + L_0$ [39.98]
[b] Measured by AFM using a diamond tip under critical normal load. The data are mean values of three tests

α is the tilt angle [39.98]. For BPT and BPTC, based on the same principle, and using the bond lengths reported in [39.99], the L_0 values are also calculated (Table 39.7). The results indicate that the measured residual thickness values of SAMs under critical load are very close to the calculated intercept length L_0 values. This means that, under the critical normal load, the Si_3N_4 tip approaches the interface, and SAMs wear away from the substrate severely. This is due to the interface chemical adsorption bond strength (S–Au) being generally smaller than the other chemical bond

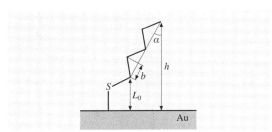

Fig. 39.17 Illustration of the relationship between the components of the equation $h = b\cos(\alpha)n + L_0$ (after [39.79])

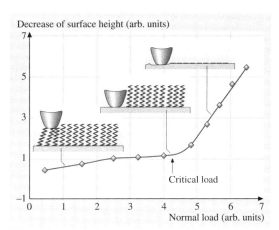

Fig. 39.18 Illustration of the wear mechanisms of SAMs with increasing normal load (after [39.81])

strengths in SAMs spacer chains (Table 39.9, to be introduced later).

According to the wear and scratch results reported here and the above discussion, the transition of the wear mechanisms of SAMs with increasing normal load is illustrated in Fig. 39.18. Below the critical normal load SAMs undergo step orientation; at the critical load SAMs wear away from the substrate due to the weak interface bond strengths; while above the critical normal load severe wear take place on the substrate. In order to improve wear resistance, the interface bond must be enhanced; a rigid spacer chain and a hard substrate are also preferred.

39.4.3 Perfluoroalkylsilane and Alkylsilane SAMs on Si(100) and Perfluoroalkylphosphonate and Alkylphosphonate SAMS on Al

Perfluorodecyltricholorosilane (PFTS)

$$CF_3-(CF_2)_7-(CH_2)_2-SiCl_3,$$

n-octyldimethyl(dimethylamino)silane (ODMS)

$$CH_3-(CH_2)_n-Si(CH_3)_2-N(CH_3)_2 \quad (n=7),$$

and n-octadecyldimethyl(dimethylamino)silane ($n = 17$) (ODDMS) vapor deposited on Si(100) substrate and perfluorodecylphosphonate (PFDP)

$$CF_3-(CF_2)_7-(CH_2)_2-\overset{\overset{\displaystyle O}{\|}}{\underset{\underset{\displaystyle O}{\|}}{P}}-OH,$$

decylphosphonate (DP)

$$\text{CH}_3-(\text{CH}_2)_n-\overset{\overset{\text{O}}{|}}{\underset{\underset{\text{O}}{\|}}{\text{P}}}-\text{OH} \quad (n=9),$$

and octadecylphosphonate (ODP) ($n = 17$) by liquid deposition on sputtered Al film on Si substrate were selected. Perfluoro-SAMs were selected because fluorinated films are known to have low surface energy. Two chain lengths of alkylsilanes (with 8 and 18 C atoms) were selected to compare their nanotribological performance with that of the former as well as to study the effect of chain length. Al substrate was selected because of the application of Al micromirrors in digital projection displays. Perfluoroalkylphosphane (with 10 C atoms) and alkylphosphonate SAMs (with 10 and 18 C atoms) on Al were selected.

Static Contact Angle and Surface Free Energy Measurements

Static contact angles and surface energy were measured as a measure of hydrophobicity. Figure 39.19 shows a Zisman plot for SAMs deposited on Si and their surface energy values using various liquid alkanes [39.59]. Critical surface tension, a measure of surface energy values for the two SAMs, are presented in the figure. Zisman analysis for the Si substrate was not available because the liquid alkanes

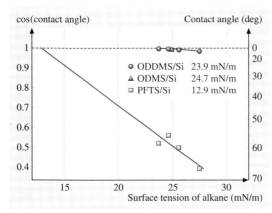

Fig. 39.19 Zisman plot for PFTS/Si, ODMS/Si, and ODDMS/Si used for calculating the critical surface tension, a measure of surface energy, which is given by the x-intercept (cos(contact angle) = 1) of the fitted line to the data

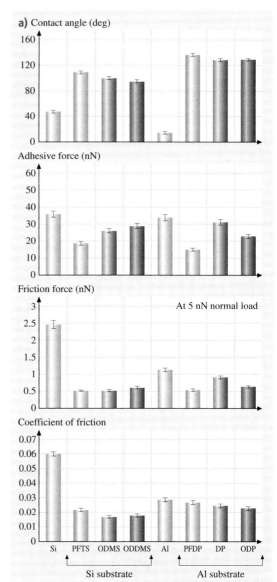

Fig. 39.20 (a) The static contact angle, adhesive force, friction force, and coefficient of friction measured using AFM for various SAMs on Si and Al substrates, and (b) friction force versus normal load plots for various SAMs on Si and Al substrates (after [39.53, 59])

used for the measurement instantly spread on these surfaces. For PFTS, significant reduction in the critical surface tension or surface energy was observed

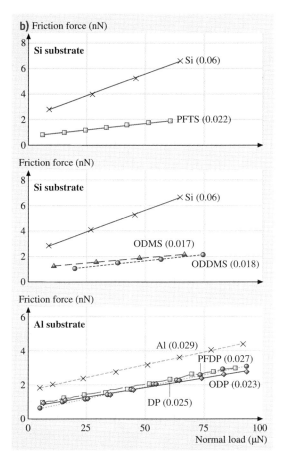

ODDMS was comparable. This suggests that the surface was comparably covered by the SAMs without bare substrate appearing.

For comparison, static contact angles were measured for various SAMs on Si and Al substrates. The measured values are compared among the samples in Fig. 39.20a [39.53, 59]. A summary of root-mean-square (RMS) roughness measured by using an AFM, and static contact angles and film thickness measured using an ellipsometer, are summarized in Table 39.8 [39.53, 61]. Significant improvement in the water-repellent property was observed for perfluorinated SAMs as compared with bare Si and Al substrates. Static contact angles of alkylsilanes and alkylphosphonates were also higher than corresponding substrates, but lower than corresponding perfluorinated films. The contact angle generally increases with a decrease in surface energy [39.104], which is consistent with the data obtained. The contact angles can be influenced by the packing density as well as the sample roughness [39.105]. The higher contact angles for the SAMs deposited on Al substrates than those on Si substrate are probably due to this effect. The $-CH_3$ groups in ODMS, ODDMS, DP, and ODP are nonpolar and are known to contribute to the water-repellent property. Perfluorinated SAMs exhibited the highest contact angle among the SAMs tested in this study.

AFM Adhesion and Friction Measurements Under Ambient Conditions

Figure 39.20a shows the adhesive force, friction force, and the coefficient of friction measured under ambient conditions using an AFM, and Fig. 39.20b shows the friction force versus normal load plots for various SAMs deposited onto Si and Al substrates [39.53, 59]. Figure 39.21 shows surface height and friction force

(12.9 mN/m for PFTS/Si) as compared with ODMS (24.7 mN/m for ODMS/Si) and ODDMS (23.9 mN/m for ODDMS/Si). The surface energy for ODMS and

Table 39.8 A summary of RMS roughness, contact angle, and film thickness of various SAMs

SAM/substrate	Acronym	RMS roughness (nm)	Contact angle (deg)	Film thickness (nm)
Silicon(111)[a]	Si	0.07	48	–
Perfluorodecyltricholorosilane/Si[a]	PFTS/Si	0.09	112	≈ 1.8
n-Octyldimethyl(dimethylamino)silane/Si[a]	ODMS/Si	0.10	99	≈ 1.9
n-Octadecyldimethyl(dimethylamino)silane/Si[b]	ODDMS/Si	0.10	92	≈ 2.1
Aluminum[b]	Al	32	< 15	–
Perfluorodecylphosphonate	PFDP	34	137	≈ 1.9
Decylphosphonate/Al[b]	DP/Al	31	129	≈ 1.9
Octadecylphosphonate/Al	ODP/Al	36	130	≈ 2.1

[a] Kasai et al. [39.61]
[b] Bhushan et al. [39.53]

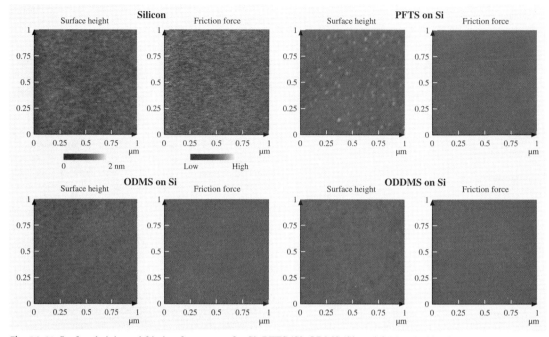

Fig. 39.21 Surface height and friction force maps for Si, PFTS/Si, ODMS/Si, and ODDMS/Si (after [39.59])

maps for Si and for PFTS, ODMS, and ODDMS on Si [39.59].

The bare substrates showed higher adhesive force than the SAMs coatings. ODMS and ODDMS show an adhesive force comparable to that of DP and ODP, despite their lower water contact angles. These SAMs have the same tail groups, and during AFM measurements the AFM tip interacts only with the tail groups, whereas the contact angles can also be influenced by the head groups in these SAMs. This is probably the reason why the adhesive forces for these SAMs are comparable. PFTS and PFDP, which have the highest contact angles, showed the lowest adhesion.

Friction force images of SAMs on Si exhibit more uniform contrast than those of bare Si. The coefficient of friction was higher for the bare substrates as compared with the corresponding SAMs deposited on them. The SAMs deposited on the Si substrate showed lower coefficient of friction than those deposited on the Al substrates. The primary reason for this is believed to be the greater roughness of the Al substrates. The SAMs with fluorocarbon backbone chains were found to have a higher coefficient of friction than those with hydrocarbon backbone chains. This might be attributed to the higher stiffness of the fluorocarbon backbone [39.53, 61]. For the fluorocarbon backbone chains, it is harder to rotate the backbone structure due to the larger size of the F atoms in comparison with the H atoms [39.106]. The C−C bonds of hydrocarbon chains can, on the other hand, rotate more freely. We presented earlier a molecular spring or brush model to explain why less compliant SAMs show larger friction. SAMs with a higher spring constant or stiffer backbone structure may need more energy to be elastically deformed during sliding; therefore friction is higher for these SAMs. In terms of the effect of chain length, it has been reported that the coefficient of friction for SAM surfaces decreases with the carbon backbone chain length (n) up to 12 C atoms ($n \approx 12$) [39.85]. This effect of chain length on the coefficient of friction was not obvious in these data.

Effect of Relative Humidity, Temperature, and Sliding Velocity on AFM Adhesion and Friction

The effect of relative humidity on adhesion and friction was studied for various SAMs. Adhesive force, friction force at 5 nN normal load, coefficient of friction, and microwear data are presented in Fig. 39.22 [39.53, 61]. The result for adhesive force for silicon showed an increase with relative humidity (Fig. 39.22a). This is ex-

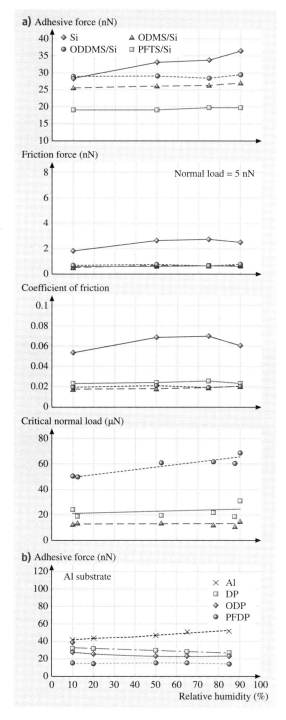

Fig. 39.22a,b Effect of relative humidity on (**a**) adhesive force, friction force, coefficient of friction, and microwear for various SAMs on (**a**) Si substrates (after [39.61]), and (**b**) adhesive force for various SAMs on Al substrates (after [39.53]) ◀

pected since the surface of silicon is hydrophilic, as shown in Fig. 39.20a. More condensation of water at the tip–sample interface at higher humidity increases the adhesive force due to the capillary effect. On the other hand, the adhesive force for the SAMs showed very weak dependency on humidity. This occurs since the surface of the SAMs is hydrophobic. The adhesive force of ODMS/Si and ODDMS/Si showed a slight increase from 75% to 90% RH. Such an increase was absent for PFTS/Si, possibly because of the hydrophobicity of PFTS/Si. The Al substrate is hydrophilic and hence shows relative humidity dependence (Fig. 39.22b). The PFDP, DP, and ODP SAMs deposited on Al substrates showed almost no change in adhesive force with humidity. The highly hydrophobic nature of these monolayers means that the contribution of water menisci to the overall adhesive force is negligible at all humidities.

The friction force of silicon showed an increase with relative humidity up to about 75% RH and a slight decrease beyond this point (Fig. 39.22a). The initial increase may result from the increase in adhesive force. The decrease in friction force at higher humidity could be attributed to the lubricating effect of the water layer. This effect is more pronounced for the coefficient of friction. Since the adhesive force increased and the coefficient of friction decreased in this range, those effects cancel each other out and the resulting friction force showed only slight changes. On the other hand, the friction force and coefficient of friction of SAMs showed very small changes with relative humidity, like that found for adhesive force. This suggests that the adsorbed water layer on the surface maintained a similar thickness throughout the range of relative humidity tested. The differences among the SAM types were small, within the measurement error, however a closer look at the coefficient of friction for ODMS/Si showed a slight increase from 75% to 90% RH as compared with PFTS/Si, possibly due to the same reason as for the adhesive force increment. The inherent hydrophobicity of SAMs means that they did not show much relative humidity dependence.

Figure 39.23a shows the effect of temperature on adhesive force, friction force at 5 nN normal load, and coefficient of friction for various SAMs on Si substrate [39.61]. Figure 39.23b shows the effect

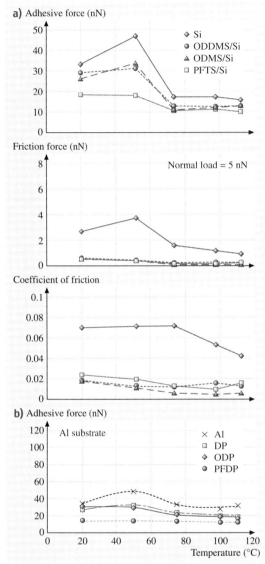

Fig. 39.23a,b Effect of temperature (**a**) on adhesive force, friction force, and coefficient of friction for various SAMs on Si substrates (after [39.61]), and (**b**) on adhesive force for various SAMs on Al substrates (after [39.53])

of temperature on the adhesive force for SAMs on Al [39.53]. The adhesive force for silicon showed an increase with temperature, from room temperature (RT) to about 55 °C, followed by a decrease from 55 °C to 75 °C, and eventually leveling off from 75 °C to 110 °C.

The adhesive force increased for Al substrate up to 50 °C and then decreased to a stable value for higher temperatures. The initial increase of adhesive force for Si and Al substrates at lower temperatures is not well understood. The observed decrease could be attributed to desorption of water molecules from the surface. After almost full depletion of the water layer, the adhesive force remains constant. The SAMs with hydrocarbon backbones on both Si and Al substrates showed similar behavior as that of the Si and Al substrates, but the initial increase in the adhesive force with temperature was smaller. The SAMs with fluorocarbon backbone chains showed almost no temperature dependence. For the SAMs with hydrocarbon backbone chains, the initial increase in adhesive force is believed to be caused by the melting of the SAM film. The melting point for a linear carbon chain molecule such as $CH_3(CH_2)_{14}CH_2OH$ is 50 °C [39.107]. With an increase in temperature, the SAM film softens, thereby increasing the real area of contact and consequently the adhesive force. Once the temperature is higher than the melting point, the lubrication regime is changed from boundary lubrication in a solid SAM to liquid lubrication in the molten SAM [39.81].

The friction force for silicon showed an increase with temperature followed by a steady decrease. The friction force is highly affected by the change in adhesion. The decrease in friction can result from the depletion of the water layer. The coefficient of friction for silicon remained constant, followed by a decrease starting at about 80 °C. For SAMs, the coefficient of friction exhibited a monotonic decrease with temperature. The decrease in friction and coefficient of friction for SAMs possibly results from the decrease in stiffness. As introduced before, the spring model suggests a smaller friction for more compliant SAMs [39.81]. The difference among the SAM types was not significant. PFTS could maintain its stiffness more than ODMS and ODDMS when temperature was increased [39.108]; however, this was not pronounced in the results.

Figure 39.24a shows the effect of sliding velocity on adhesive force, friction force, and coefficient of friction for various SAMs on a Si substrate [39.61]. The adhesive force for silicon remained rather constant at lower sliding velocity, and then increased rapidly. A similar trend was found for the SAMs. The increase in adhesive force for silicon is believed to be the result of a tribochemical reaction at the tip–sample interface [39.27] and increase of contact area by mechanical plowing. For the SAMs, the higher adhesive force at

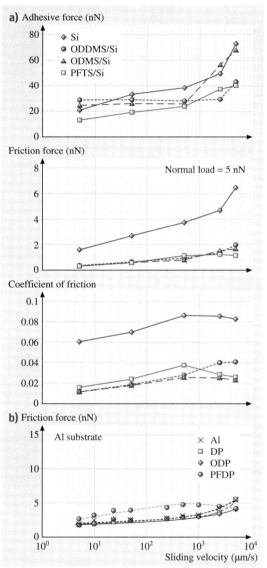

Fig. 39.24a,b Effect of sliding velocity (**a**) on adhesive force, friction force, and coefficient of friction for various SAMs on Si substrates (after [39.61]), and (**b**) on friction force for various SAMs on Al substrates (after [39.53])

the SAMs. The rate of increase was larger for ODMS than PFTS, presumably because of the higher stiffness and more densely packed structure of PFTS.

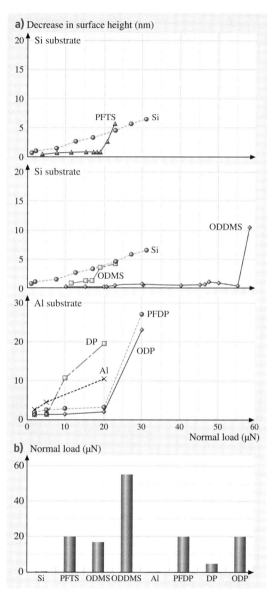

Fig. 39.25 (**a**) Decrease in surface height as a function of normal load after one scan cycle for various SAMs on Si and Al substrates, and (**b**) comparison of critical loads for failure during wear tests for various SAMs (after [39.53, 61])

the higher velocity can result from viscous drag of the SAM molecules [39.109]. SAMs can be detached from the surface and attached to an AFM tip. In addition, another reason may be an increase of contact area, which may be caused by more penetration of the AFM tip into

Table 39.9 Typical bond strengths[a] in SAMs

SAMs	Bond	HDT (kJ/mol)	BPT (kJ/mol)	Bond	PFTS (kJ/mol)	ODMS or ODDMS (kJ/mol)	PFDP (kJ/mol)	DP and ODP (kJ/mol)
Interfacial bonds	S–Au	184[b]	184[b]	Si–O	242[c]	242[c]	–	–
	S–C	286[a]	–		800[d]	800[d]	–	–
	C_6H_5–S	–	362[a]	Si–C	414[a]	414[a]	–	–
				Al–O	–	–	511[a]	511[a]
				P–C	–	–	513[a]	513[a]
				P–O	–	–	599[a]	599[a]
	C–C	–	–	C–C	–	–	–	–
Bonds in backbone	CH_2–CH_2	326[e]	–	CH_2–CH_2	326[e]	326[e]	326[e]	326[e]
	CH_3–CH_2	≈ 305[a]	–	CF_2–CF_2	≈ 326[f]	–	≈ 326[f]	–
	C_6H_5	–	Strong	CF_2–CH_2	≈ 326[f]	–	≈ 326[f]	–
				CF_3–CF_2	≈ 326[f]	–	≈ 326[f]	–
				CH_3–CH_2	–	≈ 305[a]	–	≈ 305[a] –

[a] *Lide* [39.107]
[b] Chemical adsorption bond from *Lio* et al. [39.87]
[c] Chemical adsorption bond from *Hoshino* [39.110]
[d] In diatomic molecules
[e] *Cottrell* [39.111]
[f] Due to C–C bond, should be close to that of CH_2–CH_2

The coefficient of friction showed an increase with sliding velocity and reached a plateau for Si, ODMS/Si, and ODDMS/Si. As the sliding speed is increased, the reorientation of the SAMs needs additional work, which might lead to increase in friction. For PFTS, the coefficient of friction decreased at larger sliding velocity, forming a peak. The peak structure may result from the viscoelastic property of SAMs [39.112].

Figure 39.24b shows the effect of sliding velocity on the friction force for various SAMs on Al substrate [39.53]. The friction force increases slowly at lower sliding velocities for the bare Al substrate and SAMs, followed by a rapid increase at higher sliding velocities, except for PFDP. The increase in friction force at high velocities (> 1 mm/s) is the result of asperity impacts and corresponding high frictional energy dissipation at the sliding interface for Al [39.113]. *Tambe* and *Bhushan* [39.109] extended the *molecular spring* model presented by *Bhushan* and *Liu* [39.79] to explain this velocity-dependent increase in friction force for compliant SAM molecules. Based on this model for the DP and ODP SAMs, the increase in friction force is believed to result from the reorientation of the SAM molecules under the tip load and during tip motion. The reorientation of the SAMs can act as an additional hindrance to tip motion when the AFM tip reverses during scanning and thus result in higher friction. The molecules can consequently become entangled and/or detached from the substrate and attach to the AFM tip.

AFM Wear Measurements

Figure 39.25a shows the relationship between the decrease in surface height as a function of the normal load during wear tests [39.53, 61]. As shown in the figure, the SAMs exhibit a critical normal load, beyond which the surface height decreases drastically. Figure 39.25a also shows the wear behavior of the Al and Si substrates. Unlike the SAMs, the substrates show a monotonic decrease in surface height with increasing normal load with wear initiating from the very beginning, i.e., even for low normal loads. Si (Young's modulus of elasticity $E = 130$ GPa [39.114, 115], hardness $H = 11$ GPa [39.32]) is relatively hard in comparison with Al ($E = 77$ GPa, $H = 0.41$ GPa), and hence the decrease in surface height for Al is much larger than that for Si for similar normal loads.

The critical loads corresponding to the sudden failure of SAMs are shown in Fig. 39.25b. Amongst all the SAMs, ODDMS shows the best performance in the wear tests, and this is believed to be because of the effect of the longer chain length. The fluorinated SAMs – PFTS and PFDP – show a higher critical load as compared with ODMS and DP for similar chain length. ODP shows a higher critical load as compared with DP because of its longer chain length. The mechanism of failure of compliant SAMs during wear tests was presented earlier in Fig. 39.18. It is believed that the SAMs fail mostly due to shearing of the molecule at the head group, that is, by means of shearing of the molecules off the substrate. Table 39.9 gives the bond strengths for various intermolecular bonds. The weakest bonds are at the interface, and hence failure is expected to be initiated at the interface.

To study the effect of relative humidity, wear tests were performed at various humidities. The bottom of Fig. 39.22a shows critical normal load as a function of relative humidity. The critical normal load showed weak

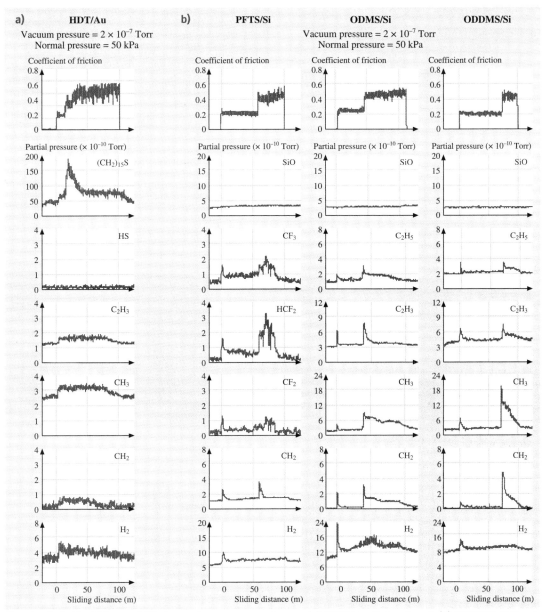

Fig. 39.26a,b Coefficients of friction and mass spectra data on (**a**) HDT/Au (1.9 nm), and (**b**) PFTS/Si (1.8 nm), ODMS/Si ($\approx$ 1.9 nm), and ODDMS/Si ($\approx$ 2.1 nm) in high vacuum (after [39.63])

dependency on relative humidity for ODMS/Si and PFTS/Si and was larger for ODMS/Si than PFTS/Si throughout the humidity range. This suggests that water molecules could penetrate into the ODDMS, which might work as a lubricant [39.81, 116]. This effect was absent for PFTS/Si and ODMS/Si.

39.4.4 Chemical Degradation and Environmental Studies

Chemical degradation and environmental studies were carried out for HDT/Au, PFTS/Si, ODMS/Si, and ODDMS/Si films.

Chemical Degradation Studies

The coefficient of friction and detected gaseous products for HDT/Au are shown in Fig. 39.26a [39.63]. A normal pressure of 50 kPa was applied on HDT films. The coefficient of friction increased after a sliding distance of about 10 m. During sliding, $(CH_2)_{15}S$, C_2H_3, CH_3, CH_2, and H_2 were detected by mass spectrometer. Partial pressure of HS fragments is of interest as it corresponds to the interface bonds, and it is reported here. Increase of $(CH_2)_{15}S$ was much more than that of other species, due to the breaking of the S—Au bond. Partial pressures of C_2H_3, CH_3, CH_2, and H_2 were also found to increase during sliding. There was no noticeable change in the partial pressure of HS.

HDT film is deposited on a Au(111) layer. The bond strength of S—Au is 184 kJ/mol (Table 39.9), which is lower than that of the C—C bond (425 kJ/mol), C—H bond (422 kJ/mol), and C—S bond (286 kJ/mol) in the alkyl chain. Since the S—Au bond is the weakest bond in the alkanethiol chain, the whole chain should be sheared away from the substrate. Because the upper atomic mass unit (amu) limit of the mass spectrometer used is 250, we monitored $(CH_2)_{15}S$ (amu = 242), which is the chain with CH_3 sheared. The rate of generation of $(CH_2)_{15}S$ is much larger than that of other species. This suggests that the mechanical shear of the whole alkanethiol chain is the dominant factor causing the failure of the HDT film. Cleavage of the S—Au bonds has been reported in literature. Based on the bond strength as well as the above studies, mechanical shearing of the C—C bonds and C—H bonds does not likely happen during sliding. The reaction induced by low-energy electrons, generated by triboelectrical emission during the sliding, could be responsible for the degradation of the alkanethiol chain. Thermal desorption of HDT from Au is another possible degradation mechanism of HDT.

The coefficient of friction and generated gaseous products for PFTS/Si, ODMS/Si, and ODDMS/Si are shown in Fig. 39.26b [39.63]. The coefficients of friction for PFTS/Si, ODMS/Si, and ODDMS/Si increase sharply after a certain sliding distance, which indicates degradation of the film. At the same time, gaseous products of CF_3, HCF_2, CF_2, CH_2, and H_2 were detected for PFTS/Si, and C_2H_5, C_2H_3, CH_3, CH_2, and H_2 were detected for ODMS/Si, and ODDMS/Si.

PFTS/Si showed lower friction than ODMS/Si in the tests. ODDMS/Si showed lower friction than both PFTS/Si and ODMS/Si. This is because of the effect of chain length. It has been reported that the coefficient of friction for SAM surfaces decreases with carbon backbone chain length (n) when the number of C atoms is less than 12. For chains with more than 12 C atoms, a change in the number of carbon atoms will not influence the coefficient of friction to a noticeable extent.

PFTS/Si showed higher durability than ODMS/Si. For the case of a perfluorinated carbon backbone, it is

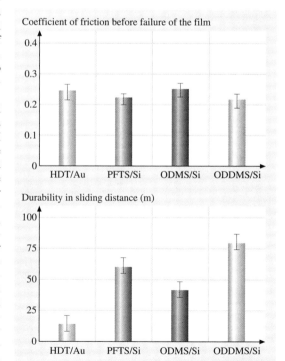

Fig. 39.27 Coefficient of friction and durability comparison of HDT/Au, PFTS/Si, ODMS/Si, and ODDMS/Si in high vacuum. *Error bars* represent ±3σ based on five measurements (after [39.63])

harder to rotate the backbone structure (due to the different size of F versus H atoms) which implies that this structure is more rigid than a hydrocarbon backbone [39.106]. *Chambers* [39.117] has reported that the C−C bond strength increases when hydrogen is replaced with fluorine. This suggests that the rigid perfluorinated carbon backbone may be responsible for the increased durability. The length of the alkyl chain also influences the desorption energy of alkanes. Based on studies of the adsorption of alkanes on Cu(100), Au(111), Pt(110), and others, the physisorption energy increases with the alkyl chain length [39.118–120]. Therefore, ODDMS is more durable than ODMS.

During sliding on PFTS films, gaseous products of CF_3, HCF_2, CF_2, CH_2, and H_2 were detected. From the structure of perfluoroalkylsilane, the only source of H on the molecular chain which could cause a partial pressure increase of H_2 is $(CH_2)_2$, which is located at the bottom of the chain. Since the partial pressure of H_2 increases immediately after sliding and remains high until the end of sliding, it is probably generated by the low-energy electrons coming from triboelectrical emission. The partial pressure of CH_2 exhibited a sharp peak at the beginning of sliding and at the moment when friction changes. Meanwhile, the partial pressures of CH_3, HSF_2, and CF_2 increased significantly when the friction increased. For ODMS and ODDMS, C_2H_5, C_2H_3, CH_3, CH_2, and H_2 were detected during sliding. The partial pressure of the carbon-related products increase considerably when the friction is increased. SiO, which is related to the interface bonds, shows no noticeable change during sliding.

Perfluoroalkylsilanes and alkylsilanes are attached to the naturally oxidized silicon by Si−O bonds. The

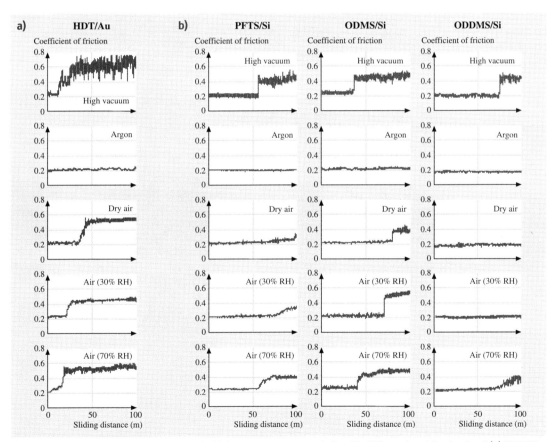

Fig. 39.28a,b Coefficient of friction data in high vacuum, argon, and air with different humidity levels of (**a**) HDT/Au (1.9 nm), and (**b**) PFTS/Si (1.8 nm), ODMS/Si ($\approx$ 1.9 nm), and ODDMS/Si ($\approx$ 2.1 nm) (after [39.63])

Si−O bond strength varies over a large range (Table 39.9), depending on the forming condition. In the alkylsilane chain, the C−Si bond strength (414 kJ/mol) is slightly lower than the C−C bond strength. Based on Table 39.9, interfacial bonds (Si−O) are weaker than the C−C bonds in the backbone. It is believed that cleavage of the films occurs at the interface. We have previously reported evidence of the cleavage of the interfacial bonds using an AFM. To explain the hydrogen, C_1 and C_2 hydrocarbon (in the tests for PFTS/Si, ODMS/Si, and ODDMS/Si) or fluorocarbon (in the tests for PFTS/Si) products, *Kluth* et al. [39.121] suggested that the alkylsilane (perfluoroalkylsilane as well) chains break and create radicals. The radical could remain on the surface and decompose to generate a shorter radical and an alkene. The radical could repeatedly decompose to ever short radicals and alkenes as long as it remains on the surface.

A summary of the coefficients of friction and durability of all the films in vacuum is presented in Fig. 39.27 [39.63].

Environmental Studies

To study the effect of environment, friction tests on HDT/Au, PFTS/Si, ODMS/Si, and ODDMS/Si were conducted in high vacuum, argon, dry air (less than 2% RH), air with 30% RH, and air with 70% RH (Fig. 39.28) [39.63]. By comparing the coefficient of friction in different environments, it was found that the friction in argon is the lowest for the SAMs tested. In high vacuum, the intimate contact leads to high friction. In dry air, the friction is higher than in the argon. This shows that oxygen has an apparent effect on the performance of SAMs. *Kim* et al. [39.122] studied the thermal stability of alkylsiloxane SAMs in air. They found that the alkylsiloxane decomposes at about 200 °C, which is much lower than the decomposition temperature of 470 °C in vacuum reported by *Kluth* et al. [39.121]. This difference could be attributed to the oxygen in air.

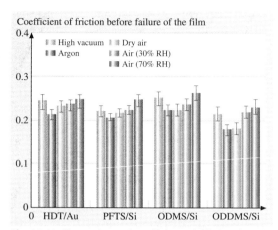

Fig. 39.29 Comparison of coefficient of friction data for HDT/Au, PFTS/Si, ODMS/Si, and ODDMS/Si in high vacuum, argon, and air with different humidity levels. *Error bars* represent ±3σ based on five measurements (after [39.63])

The water contained in air is found to have a significant influence on the friction of SAMs. A study by *Tian* et al. [39.116] on the effects of humidity on alkylsilane on mica substrate indicated that water molecules can penetrate the alkylsilane film, which alters their molecular chain ordering and can also detach alkylsilane molecules from the substrate.

A summary of the coefficients of friction before the failure of the lubricant films in various environments is presented in Fig. 39.29 [39.63]. The data in Fig. 39.29 are average values based on five measurements. To summarize the highlights, friction of the tested lubricant films is high in high vacuum because of the intimate contact between the lubricants and the counterpart surface. Friction of tested lubricant films is lower in argon than in dry air. Friction of SAMs is significantly influenced by water molecules.

39.5 Summary

Exposure of devices to a humid environment results in condensation of water vapor from the environment. Condensed water or a preexisting film of liquid forms concave meniscus bridges between the hydrophilic mating surfaces. The negative Laplace pressure present in the meniscus results in an adhesive force which depends on the interface roughness, surface tension, and contact angle. The adhesive force can be significant in an interface with ultrasmooth surfaces and it can be on the same order as the external load if the latter is small, such as in micro- and nanodevices. High adhesion also leads to stiction, friction, and wear issues in sliding surfaces. In various applications, surfaces need to be protected from exposure to the operating environment, and hy-

Table 39.10 Summary of nanotribological characterization studies for SAMs on Si and Al substrates

SAMs property		Friction force	Adhesive force	Wear
Substrate	Hard	High	Low	Low
	Soft	Low	Low	High
Chemical structure	Linear chain molecule	High	Low	High
	Ring molecule	High	High	Low
Backbone	Fluorocarbon backbone	Low	Low	Low
	Hydrocarbon backbone	Low	High	High
Chain length	Long backbone chain		High	High
	Short backbone chain		Low	Low

drophobic films are of interest. Hydrophobic films are also expected to provide low adhesion, stiction, friction, and wear. These films should be molecularly thick, well-organized, chemically bonded to the substrate, and insensitive to environment. Ordered molecular assemblies with high hydrophobicity can be engineered using chemical grafting of various polymer molecules with suitable functional head groups and nonpolar surface terminal groups.

The contact angle, adhesion, friction, and wear properties of various SAMs having alkyl, biphenyl, and perfluoroalkyl spacer chains with different surface terminal groups (−CH₃ and −CF₃) and head groups (−SH, −Si−O−, −OH, and P−O−) are presented in this chapter. It is found that the adhesive force varies linearly with the work of adhesion value of SAMs, which indicates that capillary condensation of water plays an important role in the adhesion of SAMs on the nanoscale under ambient conditions. SAMs with high-compliance long carbon spacer chains exhibit the lowest adhesive force and friction force. The friction data can be explained using a molecular spring model, in which the local stiffness and intermolecular force govern frictional performance. The results of the stiffness and friction characterization of the micropatterned sample with different structures support this model. Perfluoroalkylsilane and perfluoroalkylphosphonate SAMs exhibit lower surface energy, higher contact angle, and lower adhesive force as compared with alkylsilane and alkylphosphonate SAMs, respectively. The substrate had little effect. The coefficients of friction of various SAMs were comparable.

The influence of relative humidity on adhesion and friction of SAMs is dominated by the thickness of the adsorbed water layer. At higher humidity, water increases friction through increased adhesion by the meniscus effect in the contact zone. With an increase in temperature, in the case of Si(111), desorption of the adsorbed water layer and reduction of the surface tension of water reduces the adhesive force and friction force. A decrease in adhesion and friction with temperature was found for all films. An increase in adhesive force and friction with sliding velocity was found for all films.

PFTS/Si showed better wear resistance than ODMS/Si. ODDMS/Si showed better wear resistance than ODMS/Si due to the effect of chain length. Wear behavior of the SAMs is mostly determined by the molecule–substrate bond strengths. Similar trends were observed for films on Al substrates.

The nanotribological characterization studies of SAMs deposited on Si and Al substrates are summarized in Table 39.10 [39.52]. SAMs deposited on Si and Al substrates show low friction and low adhesion, both of which are desirable for MEMS/NEMS applications.

Based on studies in high vacuum (2×10^{-7} Torr), the coefficients of friction of the SAMs are in the following order (from low to high): ODDMS/Si, PFTS/Si, HDT/Au, ODMS/Si. HDT on Au shows lower durability than perfluoroalkylsilane and alkylsilane on Si because of the weak interfacial bond. PFTS/Si has better durability than ODMS/Si. This indicates that fluorination of alkylsilane can improve durability. ODDMS/Si is more durable than ODMS/Si and PFTS/Si because of the effect of chain length. The friction of SAMs in high vacuum is higher than in argon because of intimate contact. Based on studies in argon and air with various relative humidities, oxygen can increase the friction and decrease the durability of SAMs. Water molecules can detach SAM molecules from the substrate, resulting in high friction and low durability.

In summary, based on the contact angle, adhesion, friction, and wear measurements of SAM films by AFM, they exhibit attractive hydrophobic and tribological properties. Fluorinated SAMs appear to exhibit superior performance. SAM films should find many applications, including in micro- and nanodevices requiring good nanotribological properties and surface protection.

References

39.1 B. Bhushan, J.N. Israelachvili, U. Landman: Nanotribology: Friction, wear and lubrication at the atomic scale, Nature **374**, 607–616 (1995)

39.2 B. Bhushan: *Tribology and Mechanics of Magnetic Storage Devices*, 2nd edn. (Springer, New York 1996)

39.3 B. Bhushan (Ed.): *Tribology Issues and Opportunities in MEMS* (Kluwer Academic, Dordrecht 1998)

39.4 B. Bhushan (Ed.): *Handbook of Micro/Nanotribology*, 2nd edn. (CRC, Boca Raton 1999)

39.5 B. Bhushan: Nanotribology and nanomechanics of MEMS/NEMS and BioMEMS/BioNEMS materials and devices, Microelectron. Eng. **84**, 387–412 (2007)

39.6 B. Bhushan (Ed.): *Nanotribology and Nanomechanics – An Introduction*, 2nd edn. (Springer, Berlin, Heidelberg 2008)

39.7 B. Bhushan: Nanotribology and nanomechanics in nano/biotechnology, Philos. Trans. R. Soc. Lond. Ser. A **366**, 1499–1537 (2008)

39.8 K.F. Man, B.H. Stark, R. Ramesham: *A Resource Handbook for MEMS Reliability* (Jet Propulsion Laboratory Press, Pasadena 1998), Rev. A

39.9 K.F. Man: *MEMS Reliability for Space Applications by Elimination of Potential Failure Modes Through Testing and Analysis* (Jet Propulsion Laboratory Press, Pasadena 2002)

39.10 D.M. Tanner, N.F. Smith, L.W. Irwin, W.P. Eaton, K.S. Helgesen, J.J. Clement, W.M. Miller, J.A. Walraven, K.A. Peterson, P. Tangyunyong, M.T. Dugger, S.L. Miller: *MEMS Reliability: Infrastructure, Test Structure, Experiments, and Failure Modes* (Sandia National Laboratories, Albuquerque 2000), SAND2000-0091

39.11 A.W. Adamson: *Physical Chemistry of Surfaces*, 5th edn. (Wiley, New York 1990)

39.12 J.N. Israelachvili: *Intermolecular and Surface Forces*, 2nd edn. (Academic, London 1992)

39.13 M.E. Schrader, G.I. Loeb (Ed.): *Modern Approaches to Wettability* (Plenum, New York 1992)

39.14 B. Bhushan: *Principles and Applications of Tribology* (Wiley, New York 1999)

39.15 B. Bhushan: *Introduction to Tribology* (Wiley, New York 2002)

39.16 S. Cai, B. Bhushan: Meniscus and viscous forces during separation of hydrophilic and hydrophobic surfaces with liquid mediated contacts, Mater. Sci. Eng. R **61**, 78–106 (2008)

39.17 B. Bhushan: Contact mechanics of rough surfaces in tribology: Multiple asperity contact, Tribol. Lett. **4**, 1–35 (1998)

39.18 B. Bhushan (Ed.): *Modern Tribology Handbook* (CRC, Boca Raton 2001)

39.19 B. Bhushan: Adhesion and stiction: Mechanisms, measurement techniques, and methods for reduction, J. Vac. Sci. Technol. B **21**, 2262–2296 (2003)

39.20 B. Bhushan, Z. Zhao: Macro- and microscale tribological studies of molecularly-thick boundary layers of perfluoropolyether lubricants for magnetic thin-film rigid disks, J. Inf. Storage Proc. Syst. **1**, 1–21 (1999)

39.21 B. Bhushan, W. Peng: Contact mechanics of multilayered rough surfaces, Appl. Mech. Rev. **55**, 435–480 (2002)

39.22 B. Bhushan, Y.C. Jung: Wetting, adhesion, and friction of superhydrophobic and hydrophilic leaves and fabricated micro-/nanopatterned surfaces, J. Phys. D **20**, 225010-1–2250-24 (2008)

39.23 M. Nosonovsky, B. Bhushan: Roughness-induced superhydrophobicity: A way to design nonadhesive surfaces, J. Phys. D **20**, 225009-1–225009-30 (2008)

39.24 F.P. Bowden, D. Tabor: *The Friction and Lubrication of Solids, Part I* (Clarendon, Oxford 1950)

39.25 W.A. Zisman: Friction, durability and wettability properties of monomolecular films on solids. In: *Friction and Wear*, ed. by R. Davies (Elsevier, Amsterdam 1959) pp. 110–148

39.26 V.N. Koinkar, B. Bhushan: Microtribological studies of unlubricated and lubricated surfaces using atomic force/friction force microscopy, J. Vac. Sci. Technol. A **14**, 2378–2391 (1996)

39.27 H. Liu, B. Bhushan: Nanotribological characterization of molecularly-thick lubricant films for applications to MEMS/NEMS by AFM, Ultramicroscopy **97**, 321–340 (2003)

39.28 A. Ulman: *An Introduction to Ultrathin Organic Films: From Langmuir–Blodgett to Self-Assembly* (Academic, San Diego 1991)

39.29 A. Ulman: Formation and structure of self-assembled monolayers, Chem. Rev. **96**, 1533–1554 (1996)

39.30 H. Hansma, F. Motamedi, P. Smith, P. Hansma, J.C. Wittman: Molecular resolution of thin, highly oriented poly (tetrafluoroethylene) films with the atomic force microscope, Polym. Commun. **33**, 647–649 (1992)

39.31 L. Scandella, A. Schumacher, N. Kruse, R. Prins, E. Meyer, R. Luethi, L. Howald, M. Scherge, J.A. Schaefer: Surface modification and mechanical properties of bulk silicon. In: *Tribology Issues and Opportunities in MEMS*, ed. by B. Bhushan (Kluwer, Dordrecht 1998) pp. 529–537

39.32 B. Bhushan: Chemical, mechanical and tribological characterization of ultra-thin and hard amorphous carbon coatings as thin as 3.5 nm: Recent developments, Diam. Relat. Mater. **8**, 1985–2015 (1999)

39.33 A. Erdemir, C. Donnet: Tribology of diamond, diamond-like carbon, and related films. In: *Modern Tribology Handbook*, Vol. 2, ed. by B. Bhushan (CRC, Boca Raton 2001) pp. 871–908

39.34 V.F. Dorfman: Diamond-like nanocomposites (DLN), Thin Solid Films **212**, 267–273 (1992)

39.35 M. Grischke, K. Bewilogua, K. Trojan, H. Dimigan: Application-oriented modification of deposition process for diamond-like-carbon based coatings, Surf. Coat. Technol. **74/75**, 739–745 (1995)

39.36 R.S. Butter, D.R. Waterman, A.H. Lettington, R.T. Ramos, E.J. Fordham: Production and wetting properties of fluorinated diamond-like carbon coatings, Thin Solid Films **311**, 107–113 (1997)

39.37 M. Grischke, A. Hieke, F. Morgenweck, H. Dimigan: Variation of the wettability of DLC coatings by network modification using silicon and oxygen, Diam. Relat. Mater. **7**, 454–458 (1998)

39.38 C. Donnet, J. Fontaine, A. Grill, V. Patel, C. Jahnes, M. Belin: Wear-resistant fluorinated diamond-like carbon films, Surf. Coat. Technol. **94–95**, 531–536 (1997)

39.39 D.J. Kester, C.L. Brodbeck, I.L. Singer, A. Kyriakopoulos: Sliding wear behavior of diamond-like nanocomposite coatings, Surf. Coat. Technol. **113**, 268–273 (1999)

39.40 H. Liu, B. Bhushan: Adhesion and friction studies of microelectromechanical systems/nanoelectromechanical systems materials using a novel microtriboapparatus, J. Vac. Sci. Technol. A **21**, 1528–1538 (2003)

39.41 B. Bhushan, H. Liu, S.M. Hsu: Adhesion and friction studies of silicon and hydrophobic and low friction films and investigation of scale effects, ASME J. Tribol. **126**, 583–590 (2004)

39.42 J.A. Zasadzinski, R. Viswanathan, L. Madsen, J. Garnaes, D.K. Schwartz: Langmuir–Blodgett films, Science **263**, 1726–1733 (1994)

39.43 J. Tian, Y. Xia, G.M. Whitesides: Microcontact printing of SAMs. In: *Thin Films – Self-Assembled Monolayers of Thiols*, Vol. 24, ed. by A. Ulman (Academic, San Diego 1998) pp. 227–254

39.44 Y. Xia, G.M. Whitesides: Soft lithography, Angew. Chem. Int. Ed. **37**, 550–575 (1998)

39.45 A. Kumar, G.M. Whitesides: Features of gold having micrometer to centimeter dimensions can be formed through a combination of stamping with an elastomeric stamp and an alkanethiol ink followed by chemical etching, Appl. Phys. Lett. **63**, 2002–2004 (1993)

39.46 S.Y. Chou, P.R. Krauss, P.J. Renstrom: Imprint lithography with 25-nm resolution, Science **272**, 85–87 (1996)

39.47 Y. Xia, E. Kim, X.M. Zhao, J.A. Rogers, M. Prentiss, G.M. Whitesides: Complex optical surfaces formed by replica molding against elastomeric masters, Science **273**, 347–349 (1996)

39.48 L.J. Hornbeck: The DMD projection display chip: A MEMS-based technology, MRS Bulletin **26**, 325–328 (2001)

39.49 M.R. Douglass: Lifetime estimates and unique failure mechanisms of the digital micromirror device (DMD), 1998 Int. Reliab. Phys. Proc., presented at 36th Annu. Int. Reliab. Phys. Symp., Reno (1998)

39.50 H. Liu, B. Bhushan: Nanotribological characterization of digital micromirror devices using an atomic force microscope, Ultramicroscopy **100**, 391–412 (2004)

39.51 H. Liu, B. Bhushan: Investigation of nanotribological and nanomechanical properties of the digital micromirror device by atomic force microscope, J. Vac. Sci. Technol. A **22**, 1388–1396 (2004)

39.52 N.S. Tambe, B. Bhushan: Nanotribological characterization of self assembled monolayers deposited on silicon and aluminum substrates, Nanotechnology **16**, 1549–1558 (2005)

39.53 B. Bhushan, M. Cichomski, E. Hoque, J.A. DeRose, P. Hoffmann, H.J. Mathieu: Nanotribological characterization of perfluoroalkylphosphonate self-assembled monolayers deposited on aluminum-coated silicon substrates, Microsyst. Technol. **12**, 588–596 (2006)

39.54 E. Hoque, J.A. DeRose, P. Hoffmann, H.J. Mathieu, B. Bhushan, M. Cichomski: Phosphonate self-assembled monolayers on aluminum surfaces, J. Chem. Phys. **124**, 174710 (2006)

39.55 E. Hoque, J.A. DeRose, G. Kulik, P. Hoffmann, H.J. Mathieu, B. Bhushan: Alkylphosphonate modified aluminum oxide surfaces, J. Phys. Chem. B **110**, 10855–10861 (2006)

39.56 E. Hoque, J.A. DeRose, P. Hoffmann, B. Bhushan, H.J. Mathieu: Alkylperfluorosilane self-assembled monolayers on aluminum: A comparison with alkylphosphonate self-assembled monolayers, J. Phys. Chem. C **111**, 3956–3962 (2007)

39.57 J.A. DeRose, E. Hoque, B. Bhushan, H.J. Mathieu: Characterization of perfluorodecanote self-assembled monolayers on aluminum and comparison of stability with phosphonate and siloxy self-assembled monolayers, Surf. Sci. **602**, 1360–1367 (2008)

39.58 B. Bhushan, A.V. Kulkarni, V.N. Koinkar, M. Boehm, L. Odoni, C. Martelet, M. Belin: Microtribological characterization of self-assembled and Langmuir–Blodgett monolayers by atomic and friction force microscopy, Langmuir **11**, 3189–3198 (1995)

39.59 B. Bhushan, T. Kasai, G. Kulik, L. Barbieri, P. Hoffmann: AFM study of perfluorosilane and alkylsilane self-assembled monolayers for anti-stiction in MEMS/NEMS, Ultramicroscopy **105**, 176–188 (2005)

39.60 B. Bhushan, D. Hansford, K.K. Lee: Surface modification of silicon surfaces with vapor phase deposited ultrathin fluorosilane films for biomedical devices, J. Vac. Sci. Technol. A **24**, 1197–1202 (2006)

39.61 T. Kasai, B. Bhushan, G. Kulik, L. Barbieri, P. Hoffmann: Nanotribological study of perfluorosilane SAMs for anti-stiction and low wear, J. Vac. Sci. Technol. B **23**, 995–1003 (2005)

39.62 K.K. Lee, B. Bhushan, D. Hansford: Nanotribological characterization of perfluoropolymer thin films for bioMEMS applications, J. Vac. Sci. Technol. A **23**, 804–810 (2005)

39.63 Z. Tao, B. Bhushan: Degradation mechanisms and environmental effects on perfluoropolyether self assembled monolayers and diamond-like carbon films, Langmuir **21**, 2391–2399 (2005)

39.64 Z. Tao, B. Bhushan: Surface modification of AFM silicon probes for adhesion and wear reduction, Tribol. Lett. **21**, 1–16 (2006)

39.65 E. Hoque, J.A. DeRose, P. Hoffmann, B. Bhushan, H.J. Mathieu: Chemical stability of nonwetting, low adhesion self-assembled monolayer films formed by perfluoroalkylsilazation of copper, J. Chem. Phys. **126**, 114706 (2007)

39.66 E. Hoque, J.A. DeRose, B. Bhushan, H.J. Mathieu: Self-assembled monolayers on aluminum and copper oxide surfaces: Surface and interface characteristics, nanotribological properties, and chemical stability. In: *Applied Scanning Probe Methods IX*, NanoScience and Technology, ed. by B. Bhushan, H. Fuchs, M. Tomitori (Springer, Berlin, Heidelberg 2008) pp. 235–281

39.67 E. Hoque, J.A. DeRose, B. Bhushan, K.W. Hipps: Low adhesion, nonwetting phosphonate self-assembled monolayer films formed on copper oxide surfaces, Ultramicroscopy **109**, 1015–1022 (2009)

39.68 A. Manz, H. Becker (Eds.): *Microsystem Technology in Chemistry and Life Sciences* (Springer, Heidelberg 1998)

39.69 J. Cheng, L.J. Krica (Eds.): *Biochip Technology* (Harwood Academic, New York 2001)

39.70 M.J. Heller, A. Guttman (Eds.): *Integrated Microfabricated Biodevices* (Marcel Dekker, New York 2001)

39.71 A. van den Berg (Ed.): *Lab-on-a-Chip: Chemistry in Miniaturized Synthesis and Analysis Systems* (Elsevier, Amsterdam 2003)

39.72 D.R. Tokachichu, B. Bhushan: Bioadhesion of polymers for BioMEMS, IEEE Trans. Nanotechnol. **5**, 228–231 (2006)

39.73 B. Bhushan, M. Cichomski: Nanotribological characterization of vapor phase deposited fluorosilane self-assembled monolayers deposited on polydimethylsiloxane surfaces for biomedical micro-/nanodevices, J. Vac. Sci. Technol. A **25**, 1285–1293 (2007)

39.74 M. Hein, L.R. Best, S. Pattison, S. Arena: *Introduction to General, Organic, and Biochemistry*, 6th edn. (Brooks/Cole, Pacific Grove 1997)

39.75 J.R. Mohrig, C.N. Hammond, T.C. Morrill, D.C. Neckers: *Experimental Organic Chemistry* (Freeman, New York 1998)

39.76 A. Ulman (Ed.): *Characterization of Organic Thin Films* (Butterworth Heinemann, Boston 1995)

39.77 C. Jung, O. Dannenberger, Y. Xu, M. Buck, M. Grunze: Self-assembled monolayers from organosulfur compounds: A comparison between sulfides, disulfides, and thiols, Langmuir **14**, 1103–1107 (1998)

39.78 W. Geyer, V. Stadler, W. Eck, M. Zharnikov, A. Golzhauser, M. Grunze: Electron-induced crosslinking of aromatic self-assembled monolayers: Negative resists for nanolithography, Appl. Phys. Lett. **75**, 2401–2403 (1999)

39.79 B. Bhushan, H. Liu: Nanotribological properties and mechanisms of alkylthiol and biphenyl thiol self-assembled monolayers studied by atomic force microscopy, Phys. Rev. B **63**, 245412-1–245412-11 (2001)

39.80 H. Liu, B. Bhushan, W. Eck, V. Stadler: Investigation of the adhesion, friction, and wear properties of biphenyl thiol self-assembled monolayers by atomic force microscopy, J. Vac. Sci. Technol. A **19**, 1234–1240 (2001)

39.81 H. Liu, B. Bhushan: Investigation of nanotribological properties of alkylthiol and biphenyl thiol self-assembled monolayers, Ultramicroscopy **91**, 185–202 (2002)

39.82 S.R. Wasserman, Y.T. Tao, G.M. Whitesides: Structure and reactivity of alkylsiloxane monolayers formed by reaction of alkylchlorosilanes on silicon substrates, Langmuir **5**, 1074–1089 (1989)

39.83 J. Ruhe, V.J. Novotny, K.K. Kanazawa, T. Clarke, G.B. Street: Structure and tribological properties of ultrathin alkylsilane films chemisorbed to solid surfaces, Langmuir **9**, 2383–2388 (1993)

39.84 V. DePalma, N. Tillman: Friction and wear of self-assembled tricholosilane monolayer films on silicon, Langmuir **5**, 868–872 (1989)

39.85 M.T. McDermott, J.B.D. Green, M.D. Porter: Scanning force microscopic exploration of the lubrication capabilities of n-alkanethiolate monolayers chemisorbed at gold: Structural basis of microscopic friction and wear, Langmuir **13**, 2504–2510 (1997)

39.86 X. Xiao, J. Hu, D.H. Charych, M. Salmeron: Chain length dependence of the frictional properties of alkylsilane molecules self-assembled on mica studied by atomic force microscopy, Langmuir **12**, 235–237 (1996)

39.87 A. Lio, D.H. Charych, M. Salmeron: Comparative atomic force microscopy study of the chain length dependence of frictional properties of alkanethiol on gold and alkylsilanes on mica, J. Phys. Chem. B **101**, 3800–3805 (1997)

39.88 H. Schönherr, G.J. Vancso: Tribological properties of self-assembled monolayers of fluorocarbon and hydrocarbon thiols and disulfides on Au(111) studied by scanning force microscopy, Mater. Sci. Eng. C **8/9**, 243–249 (1999)

39.89 V.V. Tsukruk, V.N. Bliznyuk: Adhesive and friction forces between chemically modified silicon and silicon nitride surfaces, Langmuir **14**, 446–455 (1998)

39.90 V.V. Tsukruk, T. Nguyen, M. Lemieux, J. Hazel, W.H. Weber, V.V. Shevchenko, N. Klimenko, E. Sheludko: Tribological properties of modified MEMS surfaces. In: *Tribology Issues and Opportunities in MEMS*, ed. by B. Bhushan (Kluwer, Dordrecht 1998) pp. 607–614

39.91 M. Fujihira, Y. Tani, M. Furugori, U. Akiba, Y. Okabe: Chemical force microscopy of self-assembled monolayers on sputtered gold films patterned by phase separation, Ultramicroscopy **86**, 63–73 (2001)

39.92 H. Liu, B. Bhushan: Orientation and relocation of biphenyl thiol self-assembled monolayers, Ultramicroscopy **91**, 177–183 (2002)

39.93 R.J. Good, C.J.V. Oss: *Modern Approaches to Wettability – Theory and Applications* (Plenum, New York 1992)

39.94 M.H.V.C. Adão, B.J.V. Saramago, A.C. Fernandes: Estimation of the surface properties of styrene-acrylonitrile random copolymers from contact angle measurements, J. Colloid Interface Sci. **217**, 94–106 (1999)

39.95 N.S. Tambe, B. Bhushan: A new atomic force microscopy based technique for studying nanoscale friction at high sliding velocities, J. Phys. D Appl. Phys. **38**, 764–773 (2005)

39.96 B. Bhushan, J. Ruan: Tribological performance of thin film amorphous carbon overcoats for magnetic recording disks in various environments, Surf. Coat. Technol. **68/69**, 644–650 (1994)

39.97 B. Bhushan, L. Yang, C. Gao, S. Suri, R.A. Miller, B. Marchon: Friction and wear studies of magnetic thin film rigid disks with glass-ceramic, glass and aluminum-magnesium substrates, Wear **190**, 44–59 (1995)

39.98 Y.F. Miura, M. Takenga, T. Koini, M. Graupe, N. Garg, R.L. Graham, T.R. Lee: Wettability of self-assembled monolayers generated from CF_3-terminated alkanethiols on gold, Langmuir **14**, 5821–5825 (1998)

39.99 M. Ratajczak-Sitarz, A. Katrusiak, Z. Kaluski, J. Garbarczyk: 4,4′-Biphenyldithiol, Acta Crystallogr. C **43**, 2389–2391 (1987)

39.100 J.I. Siepman, I.R. McDonald: Monte Carlo simulation of the mechanical relaxation of a self-assembled monolayer, Phys. Rev. Lett. **70**, 453–456 (1993)

39.101 M. Garcia-Parajo, C. Longo, J. Servat, P. Gorostiza, F. Sanz: Nanotribological properties of octadecyltrichlorosilane self-assembled ultrathin films studied by atomic force microscopy: Contact and tapping modes, Langmuir **13**, 2333–2339 (1997)

39.102 D. DeVecchio, B. Bhushan: Localized surface elasticity measurements using an atomic force microscope, Rev. Sci. Instrum. **68**, 4498–4505 (1997)

39.103 E. Barrena, S. Kopta, D.F. Ogletree, D.H. Charych, M. Salmeron: Relationship between friction and molecular structure: Alkylsilane lubricant films under pressure, Phys. Rev. Lett. **82**, 2880–2883 (1999)

39.104 N. Eustathopoulos, M. Nicholas, B. Drevet: *Wettability at High Temperature* (Pergamon, Amsterdam 1999)

39.105 S. Ren, S. Yang, Y. Zhao, T. Yu, X. Xiao: Preparation and characterization of ultrahydrophobic surface based on a stearic acid self-assembled monolayer over polyethyleneimine thin films, Surf. Sci. **546**, 64–74 (2003)

39.106 E.S. Clark: The molecular conformations of polytetrafluoroethylene: Forms II and IV, Polymer **40**, 4659–4665 (1999)

39.107 D.R. Lide: *CRC Handbook of Chemistry and Physics*, 85th edn. (CRC, Boca Raton 2004)

39.108 W.D. Callister: *Materials Science and Engineering*, 4th edn. (Wiley, New York 1997)

39.109 N.S. Tambe, B. Bhushan: Friction model for velocity dependence of nanoscale friction, Nanotechnology **16**, 2309–2324 (2005)

39.110 T. Hoshino: Adsorption of atomic and molecular oxygen and desorption of silicon monoxide on Si(111) surfaces, Phys. Rev. B **59**, 2332–2340 (1999)

39.111 T.L. Cottrell: *The Strength of Chemical Bonds*, 2nd edn. (Butterworth, London 1958)

39.112 S.C. Clear, P.F. Nealey: The effect of chain density on the frictional behavior of surfaces modified with alkylsilanes and immersed in n-alcohols, J. Chem. Phys. **114**, 2802–2811 (2001)

39.113 N.S. Tambe, B. Bhushan: Durability studies of micro/nanoelectromechanical systems materials, coatings and lubricants at high sliding velocities (up to 10 mm/s) using a modified atomic force microscope, J. Vac. Sci. Technol. A **23**, 830–835 (2005)

39.114 Anonymous: *Properties of Silicon*, EMIS Data Rev. Ser., Vol. 4 (INSPEC Institution of Electrical Engineers, London 1988)

39.115 Anonymous: *MEMS Materials Database* (MEMS and Nanotechnology Clearinghouse, Reston 2002), http://www.memsnet.org/material/

39.116 F. Tian, X. Xiao, M.M.T. Loy, C. Wang, C. Bai: Humidity and temperature effect on frictional properties of mica and alkylsilane monolayer self-assembled on mica, Langmuir **15**, 244–249 (1999)

39.117 R.D. Chambers: *Fluorine in Organic Chemistry* (Wiley, New York 1973)

39.118 B.A. Sexton, A.E. Hughes: A comparison of weak molecular adsorption of organic-molecules on clean copper and platinum surfaces, Surf. Sci. **140**, 227–248 (1984)

39.119 L.H. Dubois, B.R. Zegarski, R.G. Nuzzo: Fundamental studies of microscopic wetting on organics surfaces, 2. Interaction of secondary adsorbates with chemically textured organic monolayers, J. Am. Chem. Soc. **112**, 570–579 (1990)

39.120 M.C. McMaster, S.L.M. Schroeder, R.J. Madix: Molecular propane adsorption dynamics on Pt(110)-(1x2), Surf. Sci. **297**, 253–271 (1993)

39.121 G.J. Kluth, M. Sander, M.M. Sung, R. Maboudian: Study of the desorption mechanism of alkylsiloxane self-assembled monolayers through isotopic labeling and high resolution electron energy-loss spectroscopy experiments, J. Vac. Sci. Technol. A **16**, 932–936 (1998)

39.122 H.K. Kim, J.P. Lee, C.R. Park, H.T. Kwak, M.M. Sung: Thermal decomposition of alkylsiloxane self-assembled monolayers in air, J. Phys. Chem. B **107**, 4348–4351 (2003)

40. Nanoscale Boundary Lubrication Studies

Bharat Bhushan

Boundary films are formed by physisorption, chemisorption, and chemical reaction. A good boundary lubricant should have a high degree of interaction between its molecules and the solid surface. As a general rule, liquids are good lubricants when they are polar and thus able to grip solid surfaces (or be adsorbed). In this chapter, we focus on various perfluoropolyethers (PFPEs) and ionic liquid films. We present a summary of nanodeformation, molecular conformation, and lubricant spreading studies, followed by an overview of the nanotribological properties of polar and nonpolar PFPEs and ionic liquid films studied by atomic force microscopy (AFM), and chemical degradation studies using a high-vacuum tribotest apparatus. In this chapter, we focus on PFPE and ionic liquid films. We first present a summary of nanodeformation, molecular conformation, and lubricant spreading studies and an overview of nanotribological and electrical studies of commonly used polar and nonpolar PFPE and ionic liquid films using AFM and chemical degradation studies using a high-vacuum tribotest apparatus.

40.1 Boundary Films 1347
40.2 Nanodeformation, Molecular Conformation, Spreading, and Nanotribological Studies 1348
    40.2.1 Nanodeformation, Molecular Conformation, and Spreading 1350
    40.2.2 Nanotribological Studies 1352
40.3 Nanotribological, Electrical, and Chemical Degradations Studies and Environmental Effects in Novel PFPE Lubricant Films 1366
    40.3.1 Nanotribological Studies 1366
    40.3.2 Wear Detection by Surface Potential Measurements 1367
    40.3.3 Wear Detection by Electrical Resistance Measurements of Z-TETRAOL and the Effect of Cycling 1369
    40.3.4 Chemical Degradation and Environmental Studies 1369
40.4 Nanotribological and Electrical Studies of Ionic Liquid Films 1375
    40.4.1 Monocationic Liquid Films 1378
    40.4.2 Dicationic Ionic Liquid Films 1383
40.5 Conclusions 1392
References 1393

40.1 Boundary Films

Boundary films are formed by physisorption, chemisorption, and chemical reaction. With physisorption, no exchange of electrons takes place between the molecules of the adsorbate and those of the adsorbant [40.1, 2]. The physisorption process typically involves van der Waals forces, which are relatively weak. In chemisorption, there is an actual sharing of electrons or an electron interchange between the chemisorbed species and the solid surface. The solid surface bonds very strongly to the adsorption species through covalent bonds. Chemically reacted films are formed by the chemical reaction of the solid surface with the environment. The physisorbed film can be either monomolecularly or polymolecularly thick. The chemisorbed films are monomolecular, but stoichiometric films formed by chemical reaction can have a large film thickness. In general, the stability and durability of surface films decrease in the following order: chemically reacted films, chemisorbed films, physisorbed films. A good boundary lubricant should have a high degree of interaction between its molecules and the solid surface. As a general rule, liquids are good lubricants when they are polar and thus able to grip solid surfaces (or be adsorbed). Polar lubricants contain reactive functional groups with low ionization potential or groups having high polarizability [40.1–3]. Boundary lubrication properties of lubricants are also dependent upon the molecular conformation and lubricant spreading [40.4–7].

Perfluoropolyether (PFPE) films, self-assembled monolayers (SAMs), and Langmuir–Blodgett (LB) films can be used as boundary lubricants [40.1–3, 8–15]. PFPE films are exclusively used for the lubrication of rigid magnetic disks and metal-evaporated magnetic tapes to reduce the friction and wear of the head–medium interface [40.10]. PFPEs are well suited for this application because of the following properties: low surface tension and high contact angle, which allow easy spreading on surfaces and provide hydrophobicity; chemical and thermal stability, which minimize degradation under use; low vapor pressure, which provides low outgassing; high adhesion to substrate via organofunctional bonds; and good lubricity, which reduces the interfacial friction and wear [40.10, 12]. While the structure of the lubricants employed at the head–medium interface has not changed substantially over the past decade, the thickness of the PFPE film used to lubricate the disk has steadily decreased from multilayer thicknesses to the sub-monolayer thickness regime [40.12, 14]. Molecularly thick PFPE films are also being considered for lubrication purposes of the evolving microelectromechanical/nanoelectromechanical systems (MEMS/NEMS) industry [40.16, 17]. It is well known that the properties of molecularly thick liquid films confined to solid surfaces can be dramatically different from those of the corresponding bulk liquid. In order to efficiently develop lubrication systems that meet the requirements of advanced rigid disk drives and MEMS/NEMS industries, the nanotribology of thin films of PFPE lubricants should be fully understood [40.11, 13, 18, 19]. It is also important to understand lubricant–substrate interfacial interactions and the influence of operating environment on the nanotribological performance of molecularly thick PFPE liquid films. Ionic liquids are considered as potential lubricants [40.20–22]. They possess efficient heat transfer properties. They are also electrically conducting, which is of interest in various MEMS/NEMS applications such as atomic force microscope probe-based data recording [40.23–25].

An overview of the nanotribological properties of SAMs and LB films can be found in many references, such as *Bhushan* [40.13]. In this chapter, we focus on PFPE and ionic liquid films. We first present a summary of nanodeformation, molecular conformation, and lubricant spreading studies and an overview of nanotribological and electrical studies of commonly used polar and nonpolar PFPE and ionic liquid films using atomic force microscopy (AFM) and chemical degradation studies using a high-vacuum tribotest apparatus.

40.2 Nanodeformation, Molecular Conformation, Spreading, and Nanotribological Studies

The molecular structures of two commonly used typical PFPE lubricants (Z-15 and Z-DOL 2000) are shown schematically in Fig. 40.1 (Fomblin, Solvay Solexis Inc., Milan, Italy). Z-15 has nonpolar −CF$_3$ end groups, whereas Z-DOL is a polar lubricant with hydroxyl (−OH) end groups. Their typical properties are summarized in Table 40.1, showing that Z-15 and Z-DOL have almost the same density and surface tension, but Z-15 has a larger molecular weight and higher viscosity. Both of them have low surface tension, low vapor pressure, low evaporation weight

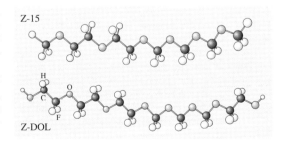

Fig. 40.1 Schematics of the molecular structures of Z-15 and Z-DOL 2000. Z-15 has nonpolar −CF$_3$ end groups, whereas Z-DOL is a polar lubricant with hydroxyl (−OH) end groups. In this figure, the m/n value, shown in Table 40.1, equals 2/3 ▶

Table 40.1 Typical properties of Z-15 and Z-DOL (Fomblin Z, Solvay Solexis Inc., Milan)

	Z-15	Z-DOL (2000)
Formula	$CF_3-O-(CF_2-CF_2-O)_m-$ $(CF_2-O)_n-CF_3$[a]	$HO-CH_2-CF_2-O-(CF_2-CF_2-O)_m-$ $(CF_2-O)_n-CF_2-CH_2-OH$[a]
Molecular weight (Da)	9100	2000
Density (ASTM D891) (g/cm^3)		
20 °C	1.84	1.81
Kinematic viscosity (ASTM D445) (cSt)		
20 °C	148	85
38 °C	90	34
99 °C	25	–
Viscosity index (ASTM D2270)	320	–
Surface tension (ASTM D1331) (dyn/cm)		
20 °C	24	24
Vapor pressure (Torr)		
20 °C	1.6×10^{-6}	2×10^{-5}
100 °C	1.7×10^{-5}	6×10^{-4}
Pour point (ASTM D972) (°C)	−80	–
Evaporation weight loss (ASTM D972) (%)		
149 °C, 22 h	0.7	–
Oxidative stability (°C)	–	320
Specific heat (cal/(g K))		
38 °C	0.21	–

[a] $m/n \approx 2/3$

loss, and good oxidative stability [40.3, 10]. For nanotribological characterization, a single-crystal Si(100) wafer with native oxide layer has generally been used as a substrate for deposition of molecularly thick lubricant films. Z-15 and Z-DOL films can be directly deposited on the Si(100) wafer by a dip-coating technique. The clean silicon wafer is submerged vertically into a dilute solution of lubricant in a hydrocarbon solvent (HT-70, Solvay Solexis Inc., Milan, Italy). The silicon wafers are vertically pulled up from the solution with a motorized stage at constant speed for deposition of desired thicknesses of Z-15 and Z-DOL lubricants [40.12, 14, 26]. The lubricant film thickness obtained in dip-coating is a function of concentration and pulling-up speed, among other factors. The Z-DOL film is bonded to the silicon substrate by heating the as-deposited Z-DOL samples in an oven at 150 °C for ≈ 30 min. The native oxide layer of Si(100) wafer reacts with the −OH groups of the lubricants during thermal treatment [40.26–29]. Subsequently, fluorocarbon solvent (FC-72, 3M) washing of the thermally treated specimen removes loosely absorbed species, leaving the chemically bonded phase on the substrate. The chemical bonding between Z-DOL molecules and the silicon substrate is illustrated in Fig. 40.2. The bonded and washed Z-DOL film is referred to as Z-DOL (fully bonded) in this chapter. The as-deposited Z-15 and Z-DOL films are mobile-phase (i.e., liquid-like) lubricants, whereas the Z-DOL (fully bonded)

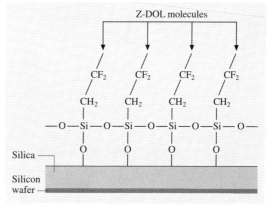

Fig. 40.2 Schematic of Z-DOL molecules chemically bonded onto a Si(100) substrate surface (which has native oxide) after thermal treatment at 150 °C for 30 min

films are fully bonded soft solid-phase (i.e., solid-like) lubricants.

40.2.1 Nanodeformation, Molecular Conformation, and Spreading

Nanodeformation behavior of Z-DOL lubricants was studied using an AFM by *Blackman* et al. [40.30, 31]. Before bringing a tungsten tip into contact with a molecular overlayer, it was first brought into contact with a bare clean silicon surface (Fig. 40.3). As the sample approaches the tip, the force is initially zero, but at point A the force suddenly becomes attractive (top curve) which increases until point B, where the sample and tip come into intimate contact and the force becomes repulsive. As the sample is retracted, a pull-off force of 5×10^{-8} N (point D) is required to overcome adhesion between the tungsten tip and the silicon surface. When an AFM tip is brought into contact with an unbonded Z-DOL film, a sudden jump in adhesive contact is also observed. A much larger pull-off force is required to overcome the adhesion. The adhesion is initiated by the formation of a lubricant meniscus surrounding the tip. This suggests that the unbonded Z-DOL lubricant shows liquid-like behavior. However, when the tip was brought into contact with a lubricant film which was firmly bonded to the surface, the liquid-like behavior disappears. The initial attractive force (point A) is no longer sudden as with the liquid film, but rather gradually increases as the tip penetrates the film.

According to *Blackman* et al. [40.30, 31], if the substrate and tip were infinitely hard with no compliance and/or deformation in the tip and sample supports, the line from B to C would be vertical with infinite slope. The tangent to the force–distance curve at a given point is referred to as the stiffness at that point and was determined by fitting a least-squares line through the nearby data points. For silicon, the deformation is reversible (elastic) since the retracting (outgoing) portion of the curve (C to D) follows the extending (ingoing) portion (B to C). For bonded lubricant film, at the point where the slope of the force changes gradually from attractive to repulsive, the stiffness changes gradually, indicating compression of the molecular film. As the load is increased, the slope of the repulsive force eventually approaches that of the bare surface. The bonded film was found to respond elastically up to the highest loads of $5\,\mu\mathrm{N}$ which could be applied. Thus, bonded lubricant behaves as a soft polymer solid.

Figure 40.4 illustrates two extremes for the conformation on a surface of a linear liquid polymer without any reactive end groups and at submonolayer coverages [40.4, 6]. At one extreme, the molecules lie flat on the surface, reaching no more than their chain diameter δ above the surface. This would be the case if a strong attractive interaction exists between the molecules and the solid. On the other extreme, when a weak attraction exists between polymer segments and the solid, the molecules adopt conformations close to that of the molecules in the bulk, with microscopic thickness equal to about the radius of gyration R_g. *Mate* and *Novotny* [40.6] used AFM to study conformation of 0.5–1.3 nm-thick Z-15 molecules on clean Si(100) surfaces. They found that the thickness measured by AFM was thicker than that measured by ellipsometry, with the offset ranging from 3 to 5 nm. They found that the offset was the same for very thin, submonolayer coverages. If the coverage is submonolayer and inadequate to make a liquid film, the relevant thickness is then the height (h_e) of the molecules extended above the solid surface. The offset should be equal to $2h_e$, assuming that the molecules extend the same height above both the tip and silicon surfaces. They thus concluded

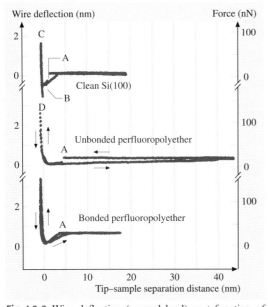

Fig. 40.3 Wire deflection (normal load) as a function of tip–sample separation distance curves comparing the behavior of clean Si(100) surface with a surface lubricated with free and unbonded PFPE lubricant, and a surface where the PFPE lubricant film was thermally bonded to the surface (after [40.30])

that the molecules do not extend > 1.5–2.5 nm above a solid or liquid surface, much smaller than the radius of gyration of the lubricants of 3.2–7.3 nm, and similar to the approximate cross-sectional diameter of 0.6–0.7 nm for the linear polymer chain. Consequently, the height that the molecules extend above the surface is considerably less than the diameter of gyration of the molecules and only a few molecular diameters in height, implying that the physisorbed molecules on a solid surface have an extended, flat conformation. They also determined the disjoining pressure of these liquid films based on AFM measurements of the distance needed to break the liquid meniscus that forms between the solid surface and an AFM tip [40.7]. For a monolayer thickness of ≈ 0.7 nm, the disjoining pressure is ≈ 5 MPa, indicating strong attractive interaction between the liquid molecules and the solid surface. The disjoining pressure decreases with increasing film thickness in a manner consistent with a strong attractive van der Waals interaction between the liquid molecules and the solid surface.

Rheological characterization shows that the flow activation energy of PFPE lubricants is weakly dependent on chain length and strongly dependent on the functional end groups [40.33]. PFPE lubricant films that contain polar end groups have lower mobility than those with nonpolar end groups of similar chain length [40.34]. The mobility of PFPE also depends on the surface chemical properties of the substrate. The spreading of Z-DOL on an amorphous carbon surface has been studied as a function of hydrogen or nitrogen content in the carbon film, using scanning microellipsometry [40.32]. The diffusion coefficient data presented in Fig. 40.5 is thickness dependent. It shows that the surface mobility of Z-DOL increased as the hydrogen content increased, but decreased as the nitrogen content increased. The enhancement of Z-DOL surface mobility by hydrogenation may be understood from the fact that the interactions between Z-DOL molecules and the carbon surface can be significantly weakened due to a reduction of the number of high-energy binding sites on the carbon surface. The stronger interactions between Z-DOL molecules and the carbon surface, as the nitrogen content in the carbon coating increases, leads to the lowering of the Z-DOL surface mobility.

Molecularly thick films may be sheared at very high shear rates, on the order of $10^8 - 10^9$ s^{-1} during sliding, such as during magnetic disk-drive operation. During such shear, lubricant stability is critical to the protection of the interface. For proper lubricant selection, viscosity at high shear rates and associated shear thinning need to

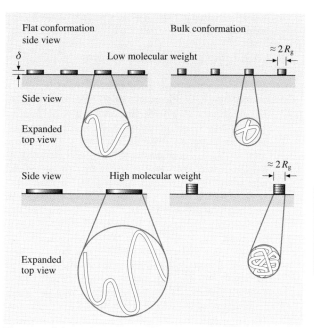

Fig. 40.4 Schematic representation of two extreme liquid conformations at the surface of the solid for low and high molecular weights at low surface coverage. δ is the cross-sectional diameter of the liquid chain and R_g is the radius of gyration of the molecules in the bulk (after [40.6])

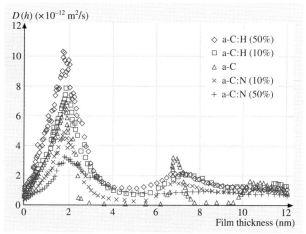

Fig. 40.5 Diffusion coefficient $D(h)$ as a function of lubricant film thickness for Z-DOL on different carbon films (after [40.32])

be understood. Viscosity measurements of eight different types of PFPE films show that all eight lubricants display Newtonian behavior and that their viscosity remains constant at shear rates up to 10^7 s^{-1} [40.35, 36].

40.2.2 Nanotribological Studies

AFM techniques have been used to investigate the nanotribological performance of PFPEs. *Mate* [40.37, 38], *O'Shea* et al. [40.39, 40], *Bhushan* et al. [40.18, 41], *Koinkar* and *Bhushan* [40.29, 42], *Bhushan* and *Sundararajan* [40.43], *Bhushan* and *Dandavate* [40.44], *Liu* and *Bhushan* [40.26], and *Palacio* and *Bhushan* [40.45, 46] used an AFM to provide insight into how PFPE lubricants function at the molecular level. *Mate* [40.37, 38] conducted friction experiments on bonded and unbonded Z-DOL, and found that the coefficient of friction of unbonded Z-DOL is about two times larger than that of bonded Z-DOL [40.39, 40]. *Koinkar* and *Bhushan* [40.29, 42] and *Liu* and *Bhushan* [40.26] studied the friction and wear performance of a Si(100) sample lubricated with Z-15, Z-DOL, and Z-DOL (fully bonded) lubricants. They found that using Z-DOL (fully bonded) could significantly improve the adhesion, friction, and wear performance of Si(100). They also discussed the lubrication mechanisms on the molecular level. *Bhushan* and *Sundararajan* [40.43] and *Bhushan* and *Dandavate* [40.44] studied the effect of tip radius and relative humidity on the adhesion and friction properties of Si(100) coated with Z-DOL (fully bonded).

In this section, we review the adhesion, friction, and wear properties of Z-15 and Z-DOL under various operating conditions (rest time, velocity, relative humidity, temperature, and tip radius). The experiments were carried out using a commercial AFM system with pyramidal Si$_3$N$_4$ and diamond tips. An environmentally controlled chamber and a thermal stage were used to perform relative humidity and temperature effect studies.

Friction and Adhesion

To investigate the friction properties of Z-15 and Z-DOL (fully bonded) films on Si(100), curves of friction force versus normal load were measured by making friction measurements at increasing normal loads [40.26]. Representative results for Si(100), Z-15, and Z-DOL (fully bonded) are shown in Fig. 40.6. An approximately linear response of all three samples is observed in the load range of 5–130 nN. The friction force of solid-like Z-DOL (fully bonded) is consistently smaller than that for Si(100), but the friction force of liquid-like Z-15 lubricant is higher than that for Si(100). *Sundararajan* and *Bhushan* [40.47] have studied the static friction force of silicon micromotors lubricated with Z-DOL by AFM. They also found that liquid-like Z-DOL lubricant significantly increased the static friction force; on the contrary a solid-like Z-DOL (fully bonded) coating can dramatically reduce the static friction force. This is in good agreement with the results of *Liu* and *Bhushan* [40.26]. In Fig. 40.6, the nonzero value of the friction force signal at zero external load is due to adhesive forces. It is well known that the following relationship exists between the friction force F and the external normal load W [40.1, 2]

$$F = \mu(W + W_a),\qquad(40.1)$$

where μ is the coefficient of friction and W_a is the adhesive force. Based on this equation and the data in Fig. 40.6, we can calculate the μ and W_a values. The coefficients of friction for Si(100), Z-15, and Z-DOL are 0.07, 0.09, and 0.04, respectively. Based on (40.1), the adhesive force values are obtained from the horizontal intercepts of the curves of friction force versus normal load, i.e., at zero friction force. Adhesive force values of Si(100), Z-15, and Z-DOL are 52, 91, and 34 nN, respectively.

The adhesive forces of these samples were also measured by using a force calibration plot (FCP) technique to obtain force–distance curves. In this technique, the tip is brought into contact with the sample, and

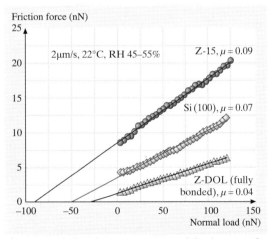

Fig. 40.6 Friction force versus normal load curves for Si(100), 2.8 nm-thick Z-15 film, and 2.3 nm-thick Z-DOL (fully bonded) film at 2 μm/s, and in ambient air sliding against a Si$_3$N$_4$ tip. Based on these curves, coefficient of friction μ and adhesion force of W_a can be calculated (after [40.26])

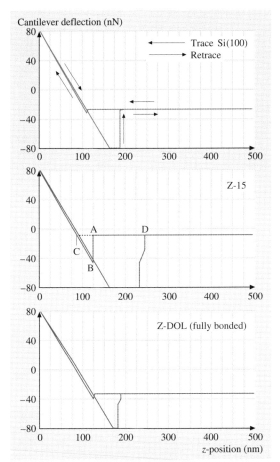

quid lubricant around the tip. From this point on, the tip is in contact with the surface, and as the piezo extends further, the cantilever is further deflected. This is represented by the slope portion of the curve. As the piezo retracts, at point C the tip goes beyond the zero-deflection (flat) line, because of the attractive forces, into the adhesive force regime. At point D, the tip snaps free of the adhesive forces and is again in free air. The adhesive force (pull-off force) is determined by multiplying the cantilever spring constant (0.58 N/m) by the horizontal distance between points C and D, which corresponds to the maximum cantilever deflection toward the samples before the tip is disengaged. Incidentally, the horizontal shift between the loading and unloading curves results from the hysteresis of the PZT tube.

The adhesive forces of Si(100), Z-15, and Z-DOL (fully bonded) measured by FCP and from plots of friction force versus normal load are summarized in Fig. 40.8 [40.26]. The results measured by these two methods are in good agreement. Figure 40.8 shows that

Fig. 40.7 Typical force calibration plots of Si(100), 2.8 nm-thick Z-15 film, and 2.3 nm-thick Z-DOL (fully bonded) film in ambient air. The adhesive forces can be calculated from the horizontal distance between points C and D, and the cantilever spring constant of 0.58 N/m (after [40.26])

the maximum force needed to pull the tip and sample apart is measured as the adhesive force. Figure 40.7 shows typical FCP curves for Si(100), Z-15, and Z-DOL (fully bonded) [40.26]. As the tip approaches the sample within a few nanometers (point A), an attractive force exists between the tip and the sample surfaces. The tip is pulled toward the sample, and contact occurs at point B on the graph. The adsorption of water molecules and/or the presence of liquid lubricant molecules on the sample surface can also accelerate this so-called *snap-in*, due to the formation of a meniscus of water and/or li-

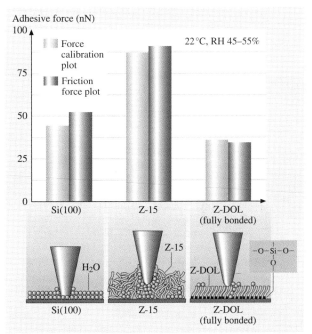

Fig. 40.8 (a) Summary of the adhesive forces of Si(100), 2.8 nm-thick Z-15 film and 2.3 nm thick Z-DOL (fully bonded) film measured by force calibration plots and plots of friction force versus normal load in ambient air. (b) Schematic showing the effect of meniscus formed between the AFM tip and the sample surface on the adhesive and friction forces (after [40.26])

the presence of a mobile Z-15 lubricant film increases the adhesive force as compared with that of Si(100). In contrast, the presence of a solid-phase Z-DOL (fully bonded) film reduces the adhesive force as compared with that of Si(100). This result is in good agreement with the results of *Blackman* et al. [40.30] and *Bhushan and Ruan* [40.48]. Sources of adhesive forces between the tip and the sample surfaces are van der Waals attraction and the long-range meniscus force [40.1, 2, 11, 13]. The relative magnitudes of the forces from these two sources are dependent upon various factors, including the distance between the tip and the sample surface, their surface roughness, their hydrophobicity, and the relative humidity [40.49]. For most surfaces with some roughness, the contribution of the meniscus force dominates at moderate to high *humidity*.

The schematic in Fig. 40.8b shows the relative size and source of the meniscus. The native oxide layer (SiO_2) on the top of the Si(100) wafer exhibits hydrophilic properties, and some water molecules can be adsorbed onto this surface. This condensed water will form a meniscus as the tip approaches the sample surface. In the case of a sphere (such as a single-asperity AFM tip) in contact with a flat surface, the attractive Laplace force F_L caused by the capillary is

$$F_L = 2\pi R \gamma (\cos\theta_1 + \cos\theta_2), \qquad (40.2)$$

where R is the radius of the sphere, γ is the surface tension of the liquid against air, and θ_1 and θ_2 are the contact angles between the liquid and flat or spherical surfaces, respectively [40.1, 2, 50]. As the surface tension value of Z-15 (24 dyn/cm) is smaller than that of water (72 dyn/cm), the larger adhesive force in Z-15 cannot only be caused by the Z-15 meniscus. The nonpolar Z-15 liquid does not have complete coverage and strong bonding with Si(100). In the ambient environment, the condensed water molecules will permeate through the liquid Z-15 lubricant film and compete with the lubricant molecules present on the substrate. The interaction of the liquid lubricant with the substrate is weakened, and a boundary layer of the liquid lubricant forms puddles [40.29, 42]. This dewetting allows water molecules to be adsorbed onto the Si(100) surface as aggregates along with Z-15 molecules, and both of them can form a meniscus while the tip approaches to the surface. In addition, as the Z-15 film is fairly soft compared with the solid Si(100) surface, and penetration of the tip in the film occurs while pushing the tip down, this leads to a large area of the tip involved to form the meniscus at the tip–liquid (water aggregates along with Z-15) interface. These two factors of the liquid-like Z-15 film result in higher adhesive force. It should also be noted that Z-15 has a higher viscosity compared with water; therefore Z-15 film provides higher resistance to sliding motion and results in a larger coefficient of friction. In the case of Z-DOL (fully bonded) film, both active groups of Z-DOL molecules are strongly bonded on the Si(100) substrate through the thermal and washing treatment, thus the Z-DOL (fully bonded) film has a relatively low free surface energy and cannot be readily displaced by water molecules or readily adsorb water molecules. Thus, the use of Z-DOL (fully bonded) can reduce the adhesive force. We further believe that the bonded Z-DOL molecules can be orientated under stress (behaving as a soft polymer solid), which facilitates sliding and reduces the coefficient of friction.

These studies suggest that liquid-like lubricant films, such as Z-15, easily form menisci (by themselves and with adsorbed water molecules), and thus have higher adhesive force and higher friction force, whereas if the lubricant film exists in a solid-like phase, such as Z-DOL (fully bonded) films, they are hydrophobic with low adhesion and friction.

In order to study the uniformity of a lubricant film and its influence on friction and adhesion, friction force mapping and adhesive force mapping of PFPE have been carried out by *Koinkar* and *Bhushan* [40.42] and *Bhushan* and *Dandavate* [40.44], respectively. Figure 40.9 shows gray scale plots of surface topography and friction force images obtained simultaneously for unbonded Demnum-type PFPE lubricant film on silicon [40.42]. Demnum-type PFPE lubricant (Demnum, Daikin, Japan) chains have $-CF_2-CH_2-OH$ (a reactive end group) on one end, whereas Z-DOL chains

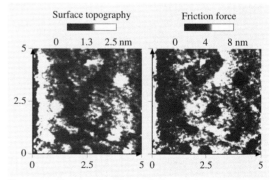

Fig. 40.9 Gray scale plots of the surface topography and friction force obtained simultaneously for unbonded 2.3 nm-thick Demnum-type PFPE lubricant film on silicon (after [40.29])

have the hydroxyl groups on both ends, as described earlier [40.12]. The friction force plot shows well-distinguished low- and high-friction regions, roughly corresponding to high- and low-surface-height regions in the topography image (thick and thin lubricant regions). A uniformly lubricated sample does not show such a variation in the friction. Figure 40.10 shows gray scale plots of the adhesive force distribution for silicon samples coated uniformly and nonuniformly with Z-DOL lubricant. It can be clearly seen that there exists a region which has an adhesive force distinctly different from the other region for the nonuniformly coated sample. This implies that the liquid film thickness is nonuniform, giving rise to a difference in the meniscus forces.

Rest-Time Effect

It is well known that, in rigid computer disk drives, the stiction force between the head and the disk magnetic medium increases rapidly with increasing rest time [40.10, 12]. Considering that stiction and friction are major issues that lead to failure of rigid computer disk drives and MEMS, it is very important to determine if this rest-time effect also exists on the nanoscale. First, the rest-time effect on the friction force, adhesive force, and coefficient of Si(100) sliding against a Si_3N_4 tip was studied (Fig. 40.11a [40.26]). It was found that the friction and adhesive forces increase logarithmically up to a certain equilibrium time, after which they remain constant. Figure 40.11a also shows that the rest time does not affect the coefficient of friction. These results suggest that the rest time can result in growth of the meniscus, which causes a higher adhesive force and in turn a higher friction force. However, over the whole testing range the friction mechanisms do not change with the rest time. Similar studies were also performed on Z-15 and Z-DOL (fully bonded) films. The results are summarized in Fig. 40.11b [40.26]. It is seen that a similar time effect was observed for Z-15 film but not for Z-DOL (fully bonded) film.

An AFM tip in contact with a flat sample surface can be treated as a single-asperity contact. Therefore, a Si_3N_4 tip in contact with Si(100) or Z-15/Si(100) can be modeled as a sphere in contact with a flat surface covered by a layer of liquid (adsorbed water and/or liquid lubricant) (Fig. 40.12a). A meniscus forms around the contacting asperity and grows with time until equilibrium occurs [40.51]. The meniscus force, which is the product of the meniscus pressure and the meniscus area, depends on the flow of liquid phase toward the contact zone. The flow of the liquid towards the contact

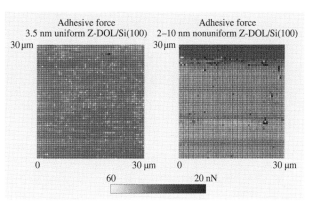

Fig. 40.10 Gray scale plots of the adhesive force distribution of a uniformly coated, 3.5 nm-thick unbonded Z-DOL film on silicon and 3–10 nm-thick unbonded Z-DOL film on silicon that was deliberately coated nonuniformly by vibrating the sample during the coating process (after [40.44])

zone is governed by the capillary pressure P_c, which draws liquid into the meniscus, and the disjoining pressure Π, which tends to draw the liquid away from the meniscus. Based on the Young–Laplace equation, the capillary pressure P_c is

$$P_c = 2\kappa\gamma , \qquad (40.3)$$

where 2κ is the mean meniscus curvature ($= \kappa_1 + \kappa_2$, where κ and κ_2 are the curvatures of the meniscus in the contact plane and perpendicular to the contact plane). *Mate* and *Novotny* [40.6] have shown that the disjoining pressure decreases rapidly with increasing liquid film thickness in a manner consistent with a strong van der Waals attraction. The disjoining pressure Π for these liquid films can be expressed as

$$\Pi = \frac{A}{6\pi h^3} , \qquad (40.4)$$

where A is the Hamaker constant and h is the liquid film thickness. The driving forces that cause the lubricant flow that result in an increase in the meniscus force are the disjoining pressure gradient due to a gradient in film thickness and the capillary pressure gradient due to the curved liquid–air interface. The driving pressure P can then be written as

$$P = -2\kappa\gamma - \Pi . \qquad (40.5)$$

Based on these three basic relationships, the following differential equation has been derived by *Chilamakuri* and *Bhushan* [40.51] to describe the meniscus

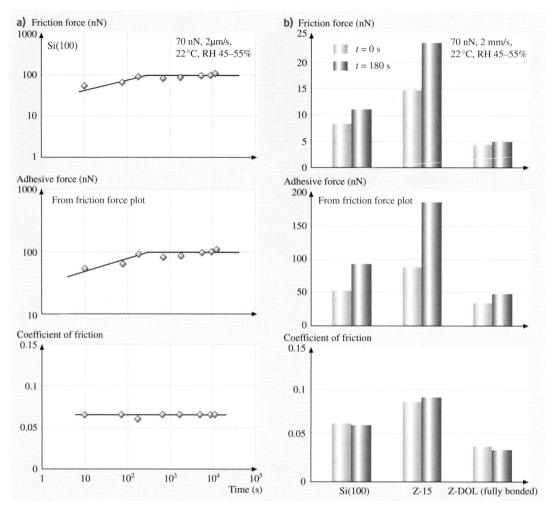

Fig. 40.11 (a) Rest-time effect on friction force, adhesive force, and coefficient of friction of Si(100). (b) Summary of the rest-time effect on friction force, adhesive force, and coefficient of friction of Si(100), 2.8 nm-thick Z-15 film, and 2.3 nm-thick Z-DOL (fully bonded) film. All of the measurements were carried out at 70 nN, 2 μm/s, and in ambient air (after [40.26])

at time t

$$2\pi x_0 \left(D + \frac{x_0^2}{2R} - h_0 \right) \frac{dx_0}{dt}$$
$$= \frac{2\pi h_0^3 \gamma}{3\eta} \frac{(1+\cos\theta)}{D+a-h_0} - \frac{Ax_0}{3\eta h} \cot\alpha , \qquad (40.6)$$

where η is the viscosity of the liquid and a is given by

$$a = R(1-\cos\phi) \propto \frac{R\phi^2}{2} \propto \frac{x_0^2}{2R} . \qquad (40.7)$$

The differential equation (40.6) was solved numerically using Newton's iteration method. The meniscus force at any time t less than the equilibrium time is proportional to the meniscus area and the meniscus pressure ($2\kappa\gamma$), and is given by

$$f_m(t) = 2\pi R\gamma(1+\cos\theta)\left[\frac{x_0}{(x_0)_{eq}}\right]^2 \left(\frac{\kappa}{\kappa_{eq}}\right) , \qquad (40.8)$$

Fig. 40.12 (a) Schematic of a single asperity in contact with a smooth flat surface in the presence of a continuous liquid film when ϕ is large. **(b)** Results of the single-asperity model. Effect of viscosity of the liquid, radius of the asperity, and film thickness is studied with respect to the time-dependent meniscus force (after [40.51]) ▶

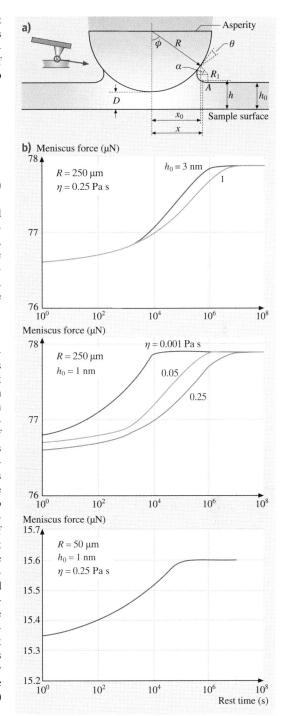

where $(x_0)_{eq}$ is the value of x_0 at the equilibrium time

$$[(x_0)_{eq}]^2 = 2R\left[\frac{-6\pi h_0^3(1+\cos\theta)}{A} + (h_0 - D)\right]. \tag{40.9}$$

This modeling work (on the microscale) showed that the meniscus force initially increases logarithmically with the rest time up to a certain equilibrium time, after which it remains constant. This equilibrium time decreases with increasing liquid film thickness, decreasing viscosity, and decreasing tip radius (Fig. 40.12b). This early numerical modeling work and the data at the nanoscale in Fig. 40.11a are in good agreement.

Velocity Effect

To investigate the effect of velocity on friction and adhesion, the friction force versus normal load relationships of Si(100), Z-15, and Z-DOL (fully bonded) at different velocities were measured (Fig. 40.13) [40.26]. Based on these data, the adhesive force and coefficient of friction values can be calculated by using (40.1). The variation of friction force, adhesive force, and coefficient of friction of Si(100), Z-15, and Z-DOL (fully bonded) as a function of velocity are summarized in Fig. 40.14, indicating that, for Si(100), the friction force decreases logarithmically with increasing velocity. For Z-15, the friction force decreases with increasing velocity up to 10 μm/s, after which it remains almost constant. Velocity has very little effect on the friction force of Z-DOL (fully bonded), which reduces slightly only at very high velocity. Figure 40.14 also indicates that the adhesive force of Si(100) is increased when the velocity is > 10 μm/s. The adhesive force of Z-15 is reduced dramatically with a velocity increase up to 20 μm/s, after which it is reduced slightly, and the adhesive force of Z-DOL (fully bonded) is also decreased at high velocity. In the tested velocity range, only the coefficient of friction of Si(100) decreases with velocity, whereas the coefficients of friction of Z-15 and Z-DOL (fully bonded) remain almost constant. This implies that the friction mechanisms of Z-15 and Z-DOL (fully bonded) do not change with velocity.

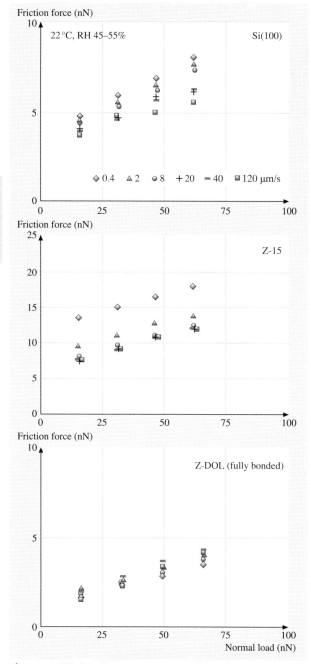

Fig. 40.13 Friction forces versus normal load data for Si(100), 2.8 nm-thick Z-15 film and 2.3 nm-thick Z-DOL (fully bonded) film at various velocities in ambient air (after [40.26])

The mechanisms of the effect of velocity on adhesion and friction can be explained based on the schematics shown in Fig. 40.14b. For Si(100), tribochemical reaction plays a major role. Although, at high velocity, the meniscus is broken and does not have enough time to rebuild, the contact stresses and high velocity lead to tribochemical reactions of the Si(100) wafer and Si_3N_4 tip, which have native oxide (SiO_2) layers, with water molecules. The following reactions occur:

$$SiO_2 + 2H_2O \rightarrow Si(OH)_4, \quad (40.10)$$
$$Si_3N_4 + 16H_2O \rightarrow 3Si(OH)_4 + 4NH_4OH. \quad (40.11)$$

The $Si(OH)_4$ is removed and continuously replenished during sliding. The $Si(OH)_4$ layer between the tip and the Si(100) surface is known to have low shear strength and causes a decrease in the friction force and coefficient of friction in the lateral direction [40.52–56]. The chemical bonds of Si–OH between the tip and the Si(100) surface induce a large adhesive force in the normal direction. For Z-15 film, at high velocity, the meniscus formed by condensed water and Z-15 molecules is broken and does not have enough time to rebuild, therefore the adhesive force and consequently the friction force is reduced. For Z-DOL (fully bonded) film, the surface can adsorb few water molecules under ambient conditions, and at high velocity these molecules are displaced, which is responsible for the slight decrease in friction force and adhesive force. Even in the high velocity range, the friction mechanisms for Z-15 and Z-DOL (fully bonded) films are still shearing of the viscous liquid and molecular orientation, respectively. Thus the coefficients of friction of Z-15 and Z-DOL (fully bonded) do not change with velocity.

Koinkar and *Bhushan* [40.29, 42] have suggested that, in the case of samples with mobile films such as condensed water and Z-15 films, alignment of liquid molecules (shear thinning) is responsible for the drop in friction force with increasing scanning velocity. This could be another reason for the decrease in friction force with velocity for Si(100) and Z-15 film in this study.

Relative Humidity and Temperature Effect

The influence of relative humidity (RH) on friction and adhesion was studied in an environmentally controlled chamber. The friction force was measured by making measurements at increasing relative humidity; the results are presented in Fig. 40.15 [40.26], which shows that, for Si(100) and Z-15 film, the friction force increases with increasing relative humidity up to RH 45%

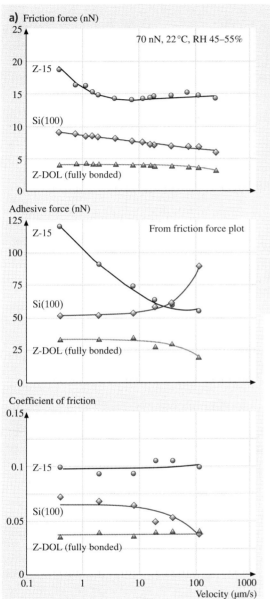

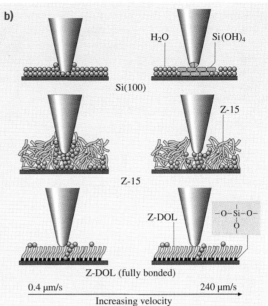

Fig. 40.14 (a) Influence of velocity on friction force, adhesive force, and coefficient of friction of Si(100), 2.8 nm-thick Z-15 film, and 2.3 nm-thick Z-DOL (fully bonded) film at 70 nN in ambient air. (b) Schematic showing the change of surface composition (by tribochemical reaction) and change of meniscus while increasing the velocity (after [40.26])

and then shows a slight decrease with further increase in relative humidity. Z-DOL (fully bonded) has a smaller friction force than Si(100) and Z-15 over the whole testing range, and its friction force shows a relative apparent increase when the relative humidity is above RH 45%. For Si(100), Z-15, and Z-DOL (fully bonded), the adhesive forces increase with relative humidity, and their coefficients of friction increase with relative humidity up to RH 45%, after which they decrease with further increase of relative humidity. It is also observed that the effect of humidity on Si(100) really depends on the history of the Si(100) sample. As the surface of Si(100) wafer readily adsorbs water from the air, without any pretreatment the Si(100) used in our study almost reaches its saturated stage of adsorbing water, which is responsible for the smaller effect with increasing relative humidity. However, if the Si(100) wafer was thermally treated by baking at 150 °C for 1 h, a larger effect was observed.

The schematic in Fig. 40.15b shows that Si(100), because of its high free surface energy, can adsorb more water molecules with increasing relative humidity. As discussed earlier, for Z-15 film in a humid environment, the condensed water from the humid environment competes with the lubricant film present on the sample surface. Obviously, more water molecules also can be

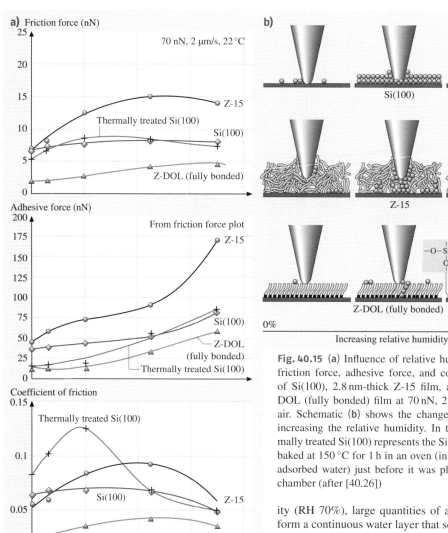

Fig. 40.15 (a) Influence of relative humidity (RH) on the friction force, adhesive force, and coefficient of friction of Si(100), 2.8 nm-thick Z-15 film, and 2.3 nm-thick Z-DOL (fully bonded) film at 70 nN, 2 μm/s, and in 22 °C air. Schematic (b) shows the change of meniscus while increasing the relative humidity. In this figure, the thermally treated Si(100) represents the Si(100) wafer that was baked at 150 °C for 1 h in an oven (in order to remove the adsorbed water) just before it was placed in the 0% RH chamber (after [40.26])

ity (RH 70%), large quantities of adsorbed water can form a continuous water layer that separates the tip and sample surface and acts as a kind of lubricant, which causes a decrease in the friction force and coefficient of friction. For Z-DOL (fully bonded) film, because of its hydrophobic surface properties, water molecules can be adsorbed and cause an increase in the adhesive force and friction force only at high humidity (RH ≥ 45%).

The effect of temperature on friction and adhesion was studied using a thermal stage attached to the AFM. The friction force was measured by making measurements at increasing temperature from 22 °C to 125 °C. The results are presented in Fig. 40.16 [40.26], which shows that the increasing temperature causes a decrease

adsorbed on Z-15 surface with increasing relative humidity. The increase in adsorbed water molecules in the case of Si(100), along with lubricant molecules in the case of Z-15 film, results in a larger water meniscus, which leads to an increase of friction force, adhesive force, and coefficient of friction of Si(100) and Z-15 with humidity. However, at a very high humid-

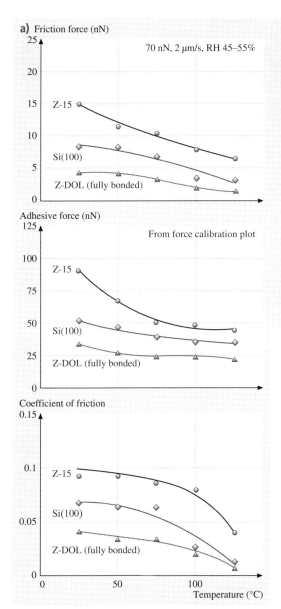
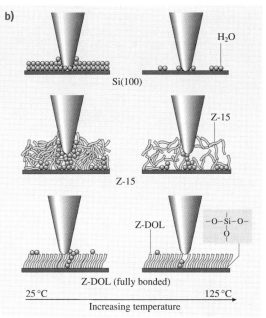

Fig. 40.16 (a) Influence of temperature on the friction force, adhesive force, and coefficient of friction of Si(100), 2.8 nm-thick Z-15 film, and 2.3 nm-thick Z-DOL (fully bonded) film at 70 nN, at $2\,\mu m/s$, and in RH 40–50% air. (b) Schematic showing that, at high temperature, desorption of water decreases the adhesive forces, and the reduced viscosity of Z-15 leads to the decrease of coefficient of friction. High temperature facilitates orientation of molecules in Z-DOL (fully bonded) film, which results in lower coefficient of friction (after [40.26])

of friction force, adhesive force, and coefficient of friction of Si(100), Z-15, and Z-DOL (fully bonded). The schematic in Fig. 40.16b indicates that, at high temperature, desorption of water leads to a decrease of friction force, adhesive force, and coefficient of friction for all of the samples. Besides that, the reduction of the surface tension of water also contributes to the decrease of friction and adhesion. For Z-15 film, the reduction of viscosity at high temperature makes an additional contribution to the decrease of friction. In the case of Z-DOL (fully bonded) film, molecules are more easily oriented at high temperature, which may also be responsible for the low friction.

Using a surface force apparatus, *Yoshizawa* and *Israelachvili* [40.57] and *Yoshizawa* et al. [40.58] have shown that a change in the velocity or temperature induces phase transformation (from crystalline solid-like, to amorphous, then to liquid-like) in surfactant monolayers, which is responsible for the observed changes in the friction force. Stick–slip is observed in the low-velocity regime of a few $\mu m/s$, and adhesion and friction first increase, followed by a decrease in the temperature range 0–50 °C. Stick–slip at low velocity, and adhesion and friction curves peaking at some particular

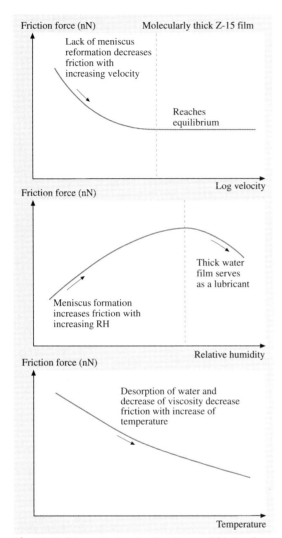

Fig. 40.17 Schematic showing the change of friction force of molecularly thick Z-15 films with log velocity, relative humidity, and temperature (after [40.26])

temperature (observed in their study), have not been observed in the AFM study. This suggests that the phase transformation may not happen in this study. This is because PFPEs generally have very good thermal stability [40.3, 10].

As a brief summary, the influence of velocity, relative humidity, and temperature on the friction force of Z-15 film is presented in Fig. 40.17. The changing trends are also addressed in this figure.

Tip-Radius Effect

The tip radius and relative humidity affect adhesion and friction for unlubricated and lubricated surfaces [40.43, 44]. Figure 40.18a shows the variation of single-point adhesive force measurements as a function of tip radius on a Si(100) sample for several humidities. The adhesive force data are also plotted as a function of relative humidity for various tip radii. Figure 40.18a indicates that the tip radius has little effect on the adhesive forces at low humidity, but the adhesive force increases with tip radius at high humidity. The adhesive force also increases with increasing humidity for all tips. The trend in adhesive forces as a function of tip radii and relative humidity (Fig. 40.18a) can be explained by the presence of meniscus forces, which arise from the capillary condensation of water vapor from the environment. If enough liquid is present to form a meniscus bridge, the meniscus force should increase with increasing tip radius based on (40.2). This observation suggests that the thickness of the liquid film at low humidity is insufficient to form continuous meniscus bridges to affect adhesive forces in the case of all tips.

Figure 40.18a also shows the variation of the coefficient of friction as a function of tip radius at a given humidity and as a function of relative humidity for a given tip radius on the Si(100) sample. It can be observed that, for RH 0%, the coefficient of friction is about the same for all the investigated tip radii except the largest one, which shows a higher value. At all other humidities, the trend consistently shows that the coefficient of friction increases with tip radius. An increase in friction with tip radius at low to moderate humidity arises from increased contact area (i.e., higher van der Waals forces) and higher values of shear forces required for the larger contact area. At high humidity, similar to the adhesive force data, an increase with tip radius occurs because of both contact area and meniscus effects. It can be seen that, for all tips, the coefficient of friction increases with humidity up to about RH 45%, beyond which it starts to decrease. This is attributed to the fact that, at higher humidity, the adsorbed water film on the surface acts as a lubricant between the two surfaces [40.26]. Thus the interface is changed at higher humidity, resulting in lower shear strength and hence lower friction force and coefficient of friction.

Figure 40.18b shows adhesive forces as a function of tip radius and relative humidity on Si(100) coated with a 0.5 nm-thick Z-DOL (fully bonded) film. Adhesive forces for all the tips with the Z-DOL (fully bonded) lubricated sample are much lower than those measured for unlubricated Si(100) (Fig. 40.18a). The

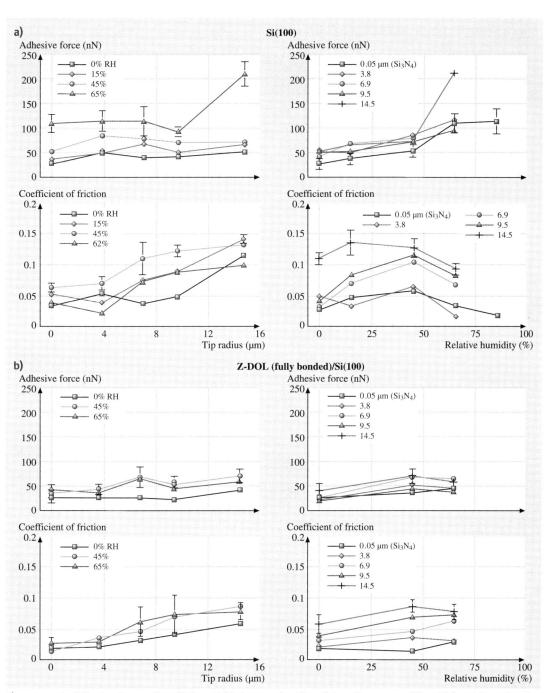

Fig. 40.18a,b Adhesive force and coefficient of friction as a function of tip radius at several humidities and as a function of relative humidity at several tip radii on (**a**) Si(100) and (**b**) 0.5 nm Z-DOL (fully bonded) films (after [40.43])

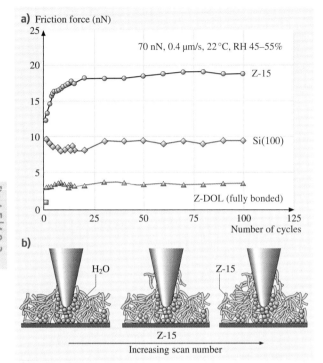

midities, as was seen on unlubricated Si(100), due to an increase in the contact area. Similar to the adhesive forces, there is an increase in friction from RH 0% to RH 45% due to a contribution from an increased number of menisci bridges. However, there is very little additional water film forming due to the hydrophobicity of the Z-DOL (fully bonded) layer thereafter, and consequentially the coefficient of friction does not change appreciably, even with the largest tip. These findings show that even a monolayer of Z-DOL (fully bonded) offers good hydrophobic performance of the surface.

Wear

To study the durability of lubricant films at the nanoscale, the friction of Si(100), Z-15, and Z-DOL (fully bonded) as a function of number of scanning cycles was measured (Fig. 40.19) [40.26]. As observed earlier, the friction force of Z-15 is higher than that of Si(100), and Z-DOL (fully bonded) has the lowest value. During cycling, the friction force of Si(100) shows a slight variation during the initial few cycles then remains constant. This is related to the removal of the top adsorbed layer. In the case of Z-15 film, the friction force shows an increase during the initial few cycles and then approaches higher and stable values. This is believed to be caused by the attachment of Z-15 molecules to the tip. The molecular interaction between these molecules attached to the tip and molecules of the film surface is responsible for the increase in friction. However, after several scans, this molecular interaction reaches equilibrium, and thereafter the fric-

Fig. 40.19 (a) Friction force versus number of sliding cycles for Si(100), 2.8 nm-thick Z-15 film, and 2.3 nm-thick Z-DOL (fully bonded) film at 70 nN, 0.8 µm/s, and in ambient air. (b) Schematic showing that some liquid Z-15 molecules can be attached onto the tip. The molecular interaction between the molecules attached to the tip and the Z-15 molecules in the film results in an increase of the friction force with multiple scans (after [40.26])

data also show that, even at a monolayer thickness of the lubricant, there is very little variation in adhesive forces with tip radius at a given humidity. For a given tip radius, the variation in adhesive forces with relative humidity indicates that these forces increase slightly from RH 0% to RH 45%, but remain more or less the same with further increase in humidity. This is seen even with the largest tip, which indicates that the lubricant is indeed hydrophobic; there is some meniscus formation at humidity above RH 0%, but it is minimal and does not increase appreciably even up to RH 65%. Figure 40.18b also shows the coefficient of friction for various tips at different humidities for the Z-DOL (fully bonded) lubricated sample. Again, all the values obtained with the lubricated sample are much lower than those obtained on unlubricated Si(100) (Fig. 40.18a). The coefficient of friction increases with tip radius for all tested hu-

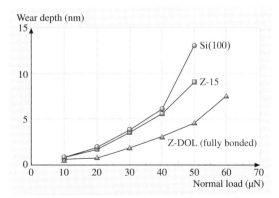

Fig. 40.20 Wear depth as a function of normal load using a diamond tip for Si(100), 2.9 nm-thick Z-15 film, and 2.3 nm-thick Z-DOL (fully bonded) after one cycle (after [40.29])

tion force and coefficient of friction remain constant. In the case of Z-DOL (fully bonded) film, the friction force starts out low and remains low during the entire test for 100 cycles. This suggests that Z-DOL (fully bonded) molecules do not become attached or displaced as readily as those of Z-15.

Koinkar and *Bhushan* [40.29, 42] conducted wear studies using a diamond tip at high loads. Figure 40.20 shows plots of wear depth as a function of normal force, and Fig. 40.21 shows wear profiles of the worn samples at 40 μN normal load. The 2.3 nm-thick Z-DOL (fully bonded) lubricated sample exhibits better wear resistance than the unlubricated and 2.9 nm-thick Z-15 lubricated silicon samples. The wear resistance of the Z-15 lubricated sample is little better than that of the

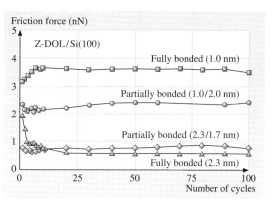

Fig. 40.22 Friction force as a function of number of cycles using a Si_3N_4 tip at a normal load of 300 nN for Z-DOL (fully bonded) and Z-DOL (partially bonded) films with different film thicknesses (after [40.42])

unlubricated sample. The Z-15 lubricated sample shows debris inside the wear track. Since Z-15 is a liquid lubricant, the debris generated is held by the lubricant, which becomes sticky and moves inside the wear track, causing damage (Fig. 40.20). These results suggest that Z-DOL (fully bonded) exhibits better wear resistance of the substrate as compared with Z-15.

To study the effect of the degree of chemical bonding, durability tests were conducted on both fully bonded and partially bonded Z-DOL films. Durability results for Z-DOL (fully bonded) and Z-DOL bonded and unwashed (partially bonded) (a partially bonded film that contains both bonded and mobile-phase lubricants) with different film thicknesses are shown in Fig. 40.22 [40.42]. Thicker films, such as Z-DOL (partially bonded) with a thickness of 4.0 nm (bonded/mobile = 2.3 nm/1.7 nm), exhibit behavior similar to that of 2.3 nm-thick Z-DOL (fully bonded) film. Figure 40.22 also indicates that Z-DOL (fully bonded) and Z-DOL (partially bonded) films with thinner film thickness exhibit higher friction values. Comparing 1.0 nm-thick Z-DOL (fully bonded) with 3.0 nm-thick (bonded/mobile = 1.0 nm/2.0 nm) Z-DOL (partially bonded), the Z-DOL (partially bonded) film exhibits lower and stable friction values. This is because the mobile phase on the surface acts as a source of lubricant replenishment. Similar conclusions have also been reported by *Ruhe* et al. [40.28], *Bhushan* and *Zhao* [40.14], and *Eapen* et al. [40.59]. All of them indicate that using partially bonded Z-DOL films can dramatically reduce friction and improve wear life.

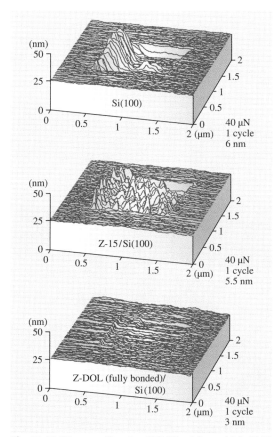

Fig. 40.21 Wear profiles for Si(100), 2.9 nm-thick Z-15 film, and 2.3 nm-thick Z-DOL (fully bonded) film after wear studies using a diamond tip. Normal force used and wear depths are listed in the figure (after [40.29])

40.3 Nanotribological, Electrical, and Chemical Degradations Studies and Environmental Effects in Novel PFPE Lubricant Films

Electrical properties of lubricant films are of interest in various MEMS/NEMS applications. Changes in the surface potential and electrical resistance can be measured during sliding using an AFM [40.20, 21, 25, 45, 46, 60, 61]. These techniques are also useful for wear detection and for studying the initiation of wear [40.19, 60]. *Palacio* and *Bhushan* [40.45, 46, 61] carried out nanotribological studies on various novel PFPE lubricant films and monitored the electrical properties as well. Chemical degradation studies and environmental effects on various PFPE lubricant films on a Si(100) wafer and magnetic tapes coated with amorphous (diamond-like) carbon were carried out in a macroscale configuration by *Tao* and *Bhushan* [40.62], *Bhushan* and *Tao* [40.63], and *Bhushan* et al. [40.64].

Structure and properties of several novel PFPE lubricants commonly used in the lubrication of magnetic rigid disks – Z-TETRAOL 2000 and A20H-2000 – are presented in Table 40.2 [40.62]. Z-DOL 2000 is also included for comparison. Z-TETRAOL (Solvay Solexis Inc.) is a derivate of PFPE. The backbone of Z-TETRAOL is the same as that of the conventional PFPE lubricant Z-DOL (Solvay Solexis Inc.) described earlier. The difference between Z-TETRAOL and Z-DOL is that Z-TETRAOL has two hydroxyl groups at each end while Z-DOL has one hydroxyl group at each end. It is believed that the two hydroxyl bonds will lead to stronger interaction with the substrate. However, the Z-TETRAOL lubricant film is less mobile because of its higher viscosity, which may lead to lower durability. A20H (Moresco, Japan) is a PFPE lubricant with a cyclotriphosphazene group at one end and a hydroxyl group at the other end. The backbone of A20H is also the same as that of Z-DOL. Phosphazene lubricants (such as X1-P) have been used as additives in the data-storage industry because they exhibit better durability in high-humidity environments [40.12, 14]. It is believed that X1-P coats the mating surface and makes it hydrophobic, minimizing stiction and improving durability. Studies have shown that A20H exhibits less thinning or reduced mobility during drive rotation [40.62]. The durability of these less mobile lubricants could be less than that of the highly mobile Z-DOL + X1-P systems. A mixture of Z-DOL and A20H is known to provide longer durability and good performance at high humidity.

The lubricants were applied on single-crystal Si(100) with a native oxide layer on the surface using the dip-coating technique by *Tao* and *Bhushan* [40.62] and *Palacio* and *Bhushan* [40.45]. The wafer was ultrasonicated in acetone followed by methanol for 10 min each. This was then followed by soaking in the solvent HFE 7100 (3M, St. Paul, MN), which consists of isomers of methoxynonafluorobutane ($C_4F_9OCH_3$). The cleaned wafer was submerged vertically into a beaker containing a dilute solution of the lubricant in HFE 7100 for 10 min and then pulled out. The lubricated sample used without post thermal treatment is referred to as untreated. Partially bonded samples were prepared by heating at 150 °C for 30 min after dip-coating, while the fully bonded samples were heated at 150 °C for 30 min and washed in HFE 7100 solvent to remove the mobile fraction. The lubricant-coated silicon samples were then measured with an ellipsometer; the coating thickness was found to be ≈ 1, 3, and 7 nm for the fully bonded, partially bonded, and untreated samples, respectively. A schematic of the bonding of the Z-DOL, Z-TETRAOL, and

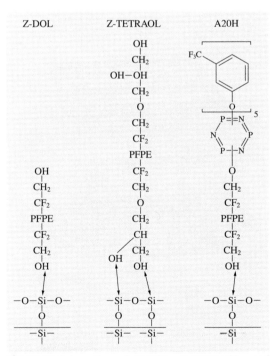

Fig. 40.23 Schematics of Z-DOL, Z-TETRAOL, and A20H molecules bonded onto Si substrate

40.3 Novel PFPE Lubricant Films

Table 40.2 Chemical structures and selected properties of several PFPE lubricants (data obtained from manufacturers' data sheet)

Lubricant	End group (X^a)	Molecular weight (amu)	Density ($\times 10^3$ kg/m^3) at 20°C	Kinematic viscosity (mm^2/s)	Surface tension (mJ/m^2)	Vapor pressure (Torr) 20°C	100°C
Z-DOL 2000^b	$-CF_2-CH_2-OH$	2000	1.81	85	24	2×10^{-5}	6×10^{-4}
Z-TETRAOL 2000^b	$-CF_2-CH_2-O-CH_2-$ $CH(OH)-CH_2-OH$	2300	1.75	2000	–	5×10^{-7}	2×10^{-4}
A20H-2000^c	$-CH_2-OH$ and $-CH_2-N_3P_3(OC_6H_4-CF_3)_5$	3000	1.7	65	22	–	–

^{a} The whole molecular chain: $X-CF_2-O-(CF_2-CF_2-O)_m-(CF_2-O)_n-CF_2-X$ ($m/n = 2/3$)
^{b} Solvay Solexis, Inc., Thorofare
^{c} Moresco Matsumura Oil Research Corp., Kobe-city, Hyogo, Japan

A20H end groups to the silicon substrate is shown in Fig. 40.23.

40.3.1 Nanotribological Studies

Adhesive force and coefficient of friction measurements were made using an AFM; the results are presented in Fig. 40.24 [40.45]. The lubricant-coated samples have reduce adhesion compared with the uncoated silicon. The adhesive forces measured on the partially bonded lubricant films are higher than the data from their fully bonded counterparts. As discussed earlier, the mobile fraction on the surface of the partially bonded sample facilitates the formation of a meniscus, which increases the tip–surface adhesion. All of the lubricant-coated samples exhibit a reduction in the coefficient of friction relative to the uncoated silicon. The partially bonded samples have a lower coefficient of friction compared with the fully bonded samples, implying that the mobile lubricant molecules in the former facilitate sliding of the tip on the surface. This will be analyzed in more detail in the following subsection in the context of wear, which was monitored using surface potential and resistance measurements.

Palacio and *Bhushan* [40.46] studied the effect of relative humidity and temperature on the friction and adhesion of novel PFPE films deposited on magnetic tapes coated with amorphous (diamond-like) carbon. The trends were similar to that reported earlier in Sect. 40.2.2.

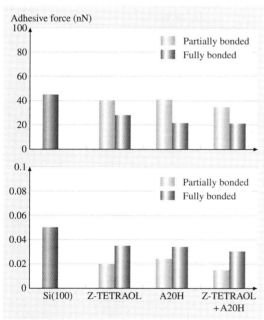

Fig. 40.24 Summary of the adhesive force and coefficient of friction of Si(100), Z-TETRAOL, A20H, and Z-TETRAOL + A20H films at room temperature (22°C) and ambient air (45–55% RH) (after [40.45])

40.3.2 Wear Detection by Surface Potential Measurements

Wear experiments were performed on the lubricated surfaces using a diamond tip over a 5×5 μm^2 region at a load of 10 μN for 20 cycles. Figure 40.25a shows surface height and surface potential images for the fully and partially bonded lubricants [40.45]. The corresponding images for the uncoated silicon substrate are shown for comparison. It can be seen from these images that debris is generated around the wear region for all coatings tested. More wear debris is observed for the fully bonded samples than for the partially bonded

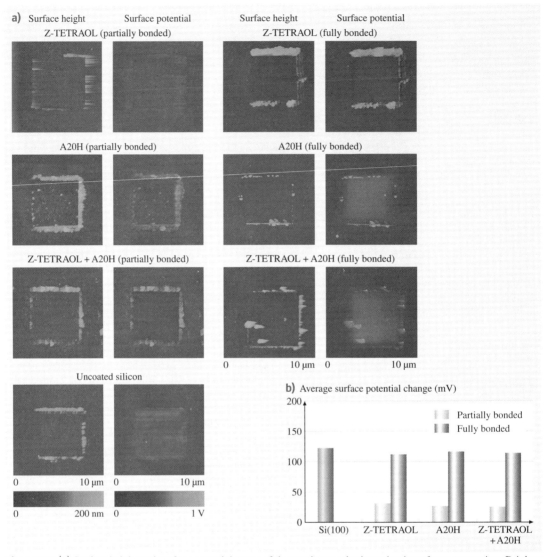

Fig. 40.25 (a) Surface height and surface potential maps of the coatings under investigation after wear testing. Brighter areas correspond to higher values of both the height and surface potential change. (b) Bar chart showing surface potential change (after [40.45])

samples. Samples with the fully bonded lubricant have limited wear protection compared with those with partially bonded lubricants because the former has only immobile molecules, while the latter has both immobile and mobile molecules. The mobile fraction of the lubricant can replenish the surface, i.e., it can move to the worn area and protect it after the immobile lubricant molecules have been displaced from the surface after repeated rubbing. Since the partially bonded films are thicker than the fully bonded films, additional wear protection is attributed to a thickness effect.

Figure 40.25b presents a bar chart showing surface potential changes for various samples. The surface potential of the area subjected to the wear test increased, an effect which is more prominent in the fully bonded samples. The partially bonded samples exhibited a smaller change in surface potential, indicating less wear. These findings should be correlated to the result of the wear test on the uncoated silicon sample, where the increase in surface potential is well understood. The Kelvin probe method measures the surface potential difference between the tip and sample, which pertains to differences in the work functions between these two materials. For conducting and semiconducting materials, the mechanism is as follows. The surface potential is altered during physical wear because the Fermi energy level is altered. This is the energy required to remove an electron to a point just outside the material surface. Thermodynamic equilibrium is disrupted with a change in the Fermi level, and can only be restored by the flow of electrons either into or from the area subjected to wear. This mechanism does not apply to materials such as SiO_2 (naturally present as a thin layer) and the lubricants, which are both insulators. Physical wear on these materials would not cause a change in the surface potential because charge dissipation is poor. Therefore, a considerable surface potential change would be observed only when:

1. The lubricant has been fully removed from the substrate;
2. The native SiO_2 layer has been abraded from the surface;
3. Wear has caused subsurface structural changes.

However, for insulators such as the lubricant, electrostatic charges are introduced as it comes into contact with a material with a dissimilar electron affinity (the diamond tip) during the wear test. The charges on the insulating lubricant surface have low mobility and would eventually dissipate into the ambient environment. In the presence of debris, the electrostatic charges become localized and may get trapped (as debris particles are mostly isolated), causing an increase in the surface potential of the debris that formed around the worn area.

40.3.3 Wear Detection by Electrical Resistance Measurements of Z-TETRAOL and the Effect of Cycling

For electrical resistance measurements, lubricant films were applied on evaporated Au film deposited on a Si substrate. The silicon wafer used in the experiments is lightly doped, and it was coated with Au so that a metal–metal contact would be attained once the lubricant was removed. PFPE is insulating. Figure 40.26 presents surface height and resistance images for a fully bonded Z-TETRAOL coating on Au by using a scanning spreading resistance microscopy (SSRM) attachment in an AFM. For comparison, another set of tests was conducted on the same sample and the post-wear surface potential images were obtained and are shown in the same figure. The Au surface is expected to have a low resistance (corresponding to darker areas in the resistance map), while the lubricant should have high resistance (lighter areas) as it is an insulator. The surface height images show a small amount of wear after one and five cycles. In both cases, the resistance images are featureless, which implies that the lubricant is still present, and the Au is not yet exposed. A surface potential increase is observed, which is attributed to electrostatic charge buildup, but this does not imply full lubricant removal since the resistance of the test area remains unchanged up to this point. After 20 cycles, the contact resistance in the tested area suddenly decreases, which indicates exposure of the underlying Au. The wear debris has the same resistance as the unworn lubricant, indicating that the current flow is influenced by the inherent difference in conductivity between the probe tip and the material it is in contact with.

During the scan of the area subjected to the wear test, the current measured by the SSRM sensor corresponds to the contact resistance between the metal-coated tip and the sample. The spreading resistance of the electric current flowing within the semiconductor sample is not measured because this is only present when there is direct contact between the metal-coated tip and a highly doped semiconductor sample.

Because the lubricant is very soft, some signal instability could occur. This comes about from possible tip contamination by the lubricant as well as plowing of wear debris during the scan. Since this AFM-based resistance measurement is a contact technique, contamination and plowing are more likely to occur compared with the surface potential measurement, which is a tapping technique. However, this technique is of interest because it provides information complementary to the Kelvin probe method in measuring the extent of wear of conducting films and lubricants, which are materials that can potentially be used in MEMS/NEMS devices.

40.3.4 Chemical Degradation and Environmental Studies

Tao and *Bhushan* [40.62] carried out chemical degradation studies in a high-vacuum tribotest apparatus [40.41,

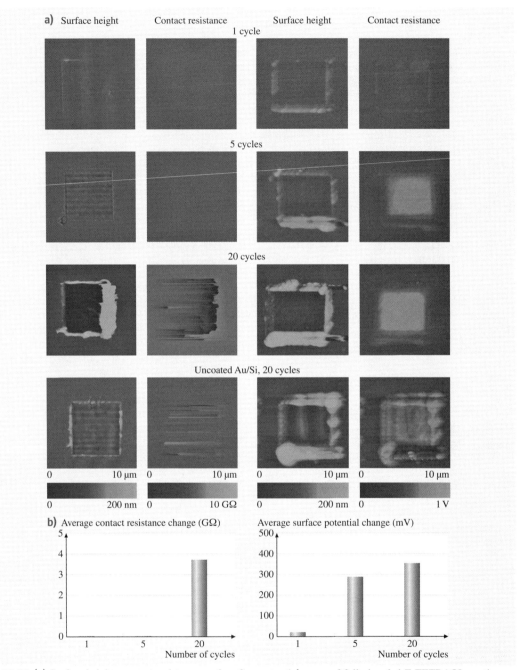

Fig. 40.26 (a) Surface height, contact resistance, and surface potential maps of fully bonded Z-TETRAOL coating on Au as a function of the number of wear cycles, and (b) bar chart showing contact resistance and surface potential change (after [40.45])

65]. In this apparatus, a lubricated wafer mounted on a flexible cantilever beam was slid against an uncoated Si(100) wafer in a macroscale configuration. The system was equipped with a mass spectrometer so that gaseous emissions from the interface could be monitored in situ during sliding in high vacuum and other controlled environments. The normal load and friction force at the contacting interface were measured using resistive-type strain-gage transducers. For the sliding tests, the lubricated Si(100) sample was glued onto a flat surface at the end of a rotating shaft. The sample was slid against a Si(100) wafer mounted on the flat surface of a slider integrated with a flexible cantilever used in magnetic rigid disk drives. The sliding speed was 0.3 m/s and the applied pressure was 150 kPa. The environmental effects were investigated in high vacuum (2×10^{-7} Torr), argon, dry air (less than 2% RH), ambient air (30% RH), and high-humidity air (70% RH).

Chemical Degradation Studies

The coefficient of friction and partial pressure of the gaseous products for fully bonded Z-DOL, Z-TETRAOL, A20H, Z-DOL + A20H (30 vol. %), and partially bonded A20H during sliding in high vacuum are shown in Fig. 40.27a–c. The result for untreated Z-DOL is also shown in the figure for comparison.

In the tests, the sharp increase of friction indicates the failure of the lubricant film. Therefore, the durability of the lubricant film can be obtained from the friction curve. Under the normal pressure of 150 kPa, the untreated Z-DOL failed immediately after sliding. The fully bonded Z-DOL began to fail after sliding of ≈ 40 m. The fully bonded Z-TETRAOL, began to fail at ≈ 80 m. Fully bonded and partially bonded A20H, however, did not fail during the 100 m sliding. The fully bonded Z-DOL + A20H (30 vol. %) failed immediately after sliding. The results show that fully bonded Z-TETRAOL and A20H are more durable than fully bonded Z-DOL, while the untreated Z-DOL is less durable than the bonded films. The PFPE lubricants can be bonded to the applied surface through the hydroxyl group. Z-TETRAOL, with two hydroxyl groups at each end of the molecular chain, can be bonded more tightly onto the silicon surface than Z-DOL. The high durability of A20H, however, is beyond expectations. To further confirm the result, a normal pressure of 200 kPa was applied on partially bonded A20H. At the normal pressure of 200 kPa, the coefficient of friction was found to increase. At the same time, gaseous products were detected (Fig. 40.27c). The final coefficient of friction was ≈ 0.3, which is lower than the values (≈ 0.4)

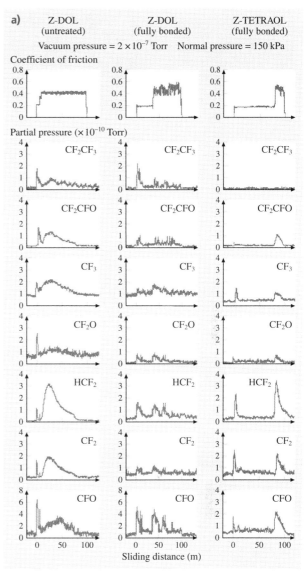

Fig. 40.27a–c Coefficients of friction and mass spectra data on (**a**) untreated (2.2 nm) and fully bonded (2.3 nm) Z-DOL, fully bonded Z-TETRAOL (2.6 nm), (**b**) fully bonded (0.9 nm) and partially bonded (2.3 nm) A20H, and Z-DOL + A20H (2.4 nm) in high vacuum and under 150 kPa normal pressure, (**c**) partially bonded A20H in high vacuum and under 200 kPa normal pressure (after [40.62]) ▲ ▶

for the other films after failure. This could indicate that the A20H film was only partially worn. The reason for

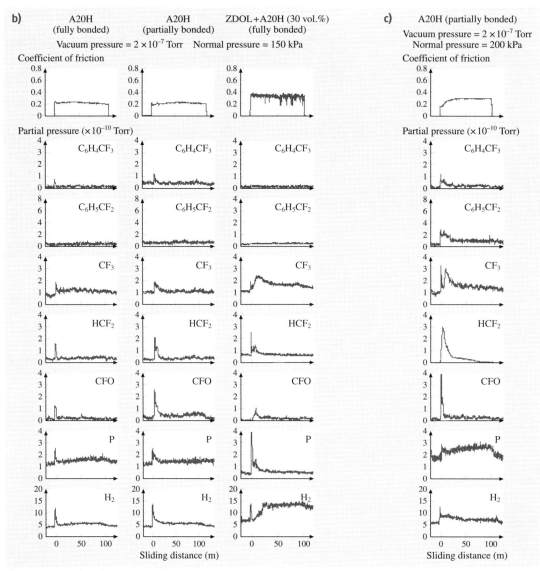

Fig. 40.27 (continued)

the lower durability of Z-DOL + A20H (30 vol. %) in high vacuum is not clear.

In the tests on Z-DOL and Z-TETRAOL films, CF_2CF_3, CF_2CFO, CF_3, CF_2O, HCF_2, CF_2, and CFO were found to increase when friction increased. In the test of A20H films, $C_6H_4CF_3$, $C_6H_5CF_2$, CF_3, HCF_2, CFO, P, and H_2 were detected. For Z-DOL + A20H (30 vol. %), $C_6H_4CF_3$, $C_6H_5CF_2$, CF_3, HCF_2, CFO, P, and H_2 were also detected during sliding. Degradation mechanisms for Z-DOL and various model lubricants have been studied by *Zhao* and *Bhushan* [40.66, 67] and *Zhao* et al. [40.68]. Based on this work, Z-DOL starts to decompose above 350 °C. In the sliding conditions used here, it is not likely that frictional heat could generate such a high temperature. The possibility of catalytic degradation is very low because of the small contact area. Triboelectrical reaction and mechanical scission are considered to be the dominant mechanisms during the sliding of Si on PFPE films. Electron emission is known to occur during sliding for both metal and nonmetal surfaces. The interaction of electrons with PFPE molecules could create various radicals such as $\bullet CF_2-O-CF_2-$, $\bullet CF_2-CF_2-O-CF_2-$, and/or $\bullet CF_2-CF_3$. These radicals can decompose or react with each other and/or the remaining PFPE molecules. A detailed description of triboelectrical reactions can be found in *Zhao* et al. [40.68] and *Zhao* and *Bhushan* [40.67].

Mechanical scission is another mechanism that can cause the degradation of PFPE films. The PFPE molecule has a long linear chain structure. The $C-C$ and $C-O$ bonds in the molecular chain are easily subjected to cleaving by microasperities on the rubbing surfaces, which results in decomposition of the lubricant during the sliding process.

A summary of the coefficients of friction and durability of the tested lubricants in high vacuum is shown in Fig. 40.28. The coefficients of friction of fully bonded Z-DOL and Z-TETRAOL are lower than those of untreated Z-DOL and A20H in high vacuum. Z-TETRAOL shows higher durability than Z-DOL, while A20H (both fully bonded and partially bonded) shows even high durability in high vacuum.

Environmental Studies

The wear tests were conducted in high vacuum, argon, dry air (< 2% RH), air with 30% RH, and air with 70% RH [40.62]. The normal pressure applied was 150 kPa for all tested films, which was the same as in the degradation tests. Test results are presented in Fig. 40.29. The coefficients of friction show that friction and durability vary with the environment. Z-DOL films fail more rapidly in high vacuum than in other environments. As described in the previous section, there are very few foreign molecules on the contacting surfaces in high vacuum. This enables intimate contact between the lubricant film and the counterpart surface. A tendency for chemical bonding occurs between the lubricant film and the counterpart surface. In argon, the Z-DOL films exhibited lower friction and higher durability than in dry air. The chemical effects of oxygen can influence the friction and durability of the lubricant films [40.65]. The water molecules at a moderate humidity level (ambient air) can act as a lubricant between the contacting surfaces. However, in a high-humidity environment, the water molecules can penetrate the Z-DOL film and result in a nonuni-

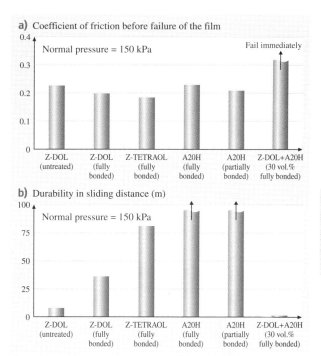

Fig. 40.28a,b Coefficient of friction (**a**) and durability comparison (**b**) of Z-DOL, Z-TETRAOL, A20H, and Z-DOL + A20H in high vacuum (after [40.62])

form distribution of the Z-DOL molecules. The fully bonded Z-DOL film is less influenced by environment than untreated Z-DOL. The reason is that the untreated Z-DOL has more free hydroxyl end groups. The water molecules in the environment can interact with the hydroxyl group of Z-DOL via hydrogen bonding.

Z-TETRAOL is more durable than Z-DOL. Especially, a high-humidity environment does not seem to have negative effects on Z-TETRAOL, although it has one more hydroxyl end group than Z-DOL and may attract more water molecules at high humidity. This could be because the two hydroxyl end groups provides better attachment than for Z-DOL on the surface. The tight bonding reduces the probability of nonuniform distribution of the film. The coefficient of friction of fully bonded Z-TETRAOL is very close to that of fully bonded Z-DOL in argon and air with various humidity levels. In high vacuum, Z-TETRAOL shows a lower coefficient of friction than Z-DOL.

Both fully bonded and partially bonded A20H were very durable in high vacuum. In argon, fully bonded

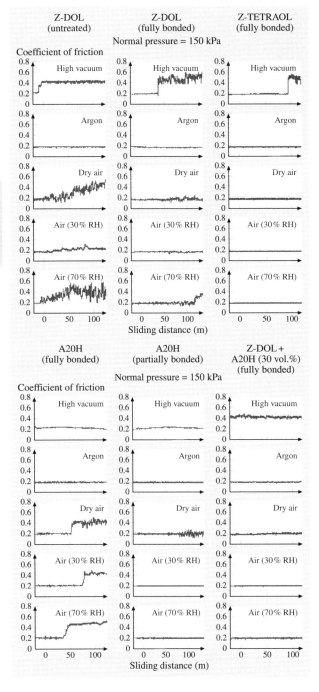

Fig. 40.29 Coefficient of friction data in high vacuum, argon, and air with different humidity levels of untreated (2.2 nm) and fully bonded (2.3 nm) Z-DOL, fully bonded Z-TETRAOL (2.6 nm), fully bonded (0.9 nm) and partially bonded (2.3 nm) A20H, and Z-DOL + A20H (2.4 nm) (after [40.62]) ◄

and partially bonded A20H did not fail during the 100 m sliding. In air with various humidity levels, however, the fully bonded A20H failed. The durability is lower compared with Z-DOL. This low durability may be caused by the low thickness of the film, which is only 0.9 nm. For the 2.3 nm-thick partially bonded A20H film, the durability in air with various humidity levels is apparently improved. Especially, at high humidity level, partially bonded A20H exhibits high durability. A20H consists of a phosphazene group that is large in size and protects the surface. Phosphazene lubricant has been used as an additive and is known to have high durability in a high-humidity environment. The coefficient of friction of A20H, however, is higher (5–15%) than that of Z-DOL in various environments.

Z-DOL + A20H (30 vol. %) shows low durability in high vacuum. In argon and air with various humidity levels, however, the film did not fail during the 100 m sliding. The durability of the mixture is higher than that of fully bonded Z-DOL. The coefficient of friction of the mixture is higher than that of Z-DOL while lower than that of A20H.

A summary of the coefficients of friction before the failure of the lubricant films in various environments is presented in Fig. 40.30. To summarize the highlights, Z-TETRAOL exhibits higher durability than Z-DOL. A20H exhibits high durability in high vacuum, in argon, and in air with various humidity levels. The mixture Z-DOL + A20H (30 vol. %) shows low durability in high vacuum but high durability in argon and air with various humidity levels.

In order to investigate the durability of the lubricant films further, a single-crystal Si(100) ball (1 mm in diameter, 5×10^{-7} atoms/cm^3 boron doped) was used as a slider on the films to accelerate wear. Tests were performed in argon, dry air, ambient air, and high-humidity air. The applied load on the silicon ball was 2.5 g, and the sliding speed was 0.3 m/s. The sliding distance was up to 600 m. The results are shown in Fig. 40.31. From Fig. 40.31, in all environments, Z-TETRAOL exhibits higher durability than the other films. The durability of A20H is slightly lower than

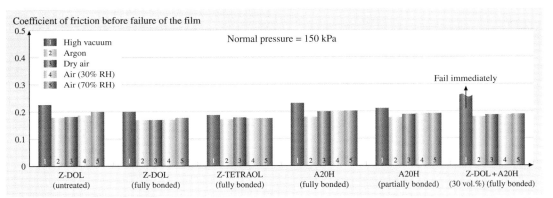

Fig. 40.30 Comparison of coefficient of friction data for Z-DOL, Z-TETRAOL, A20H, and Z-DOL + A20H in high vacuum, argon, and air with different humidity levels (after [40.62])

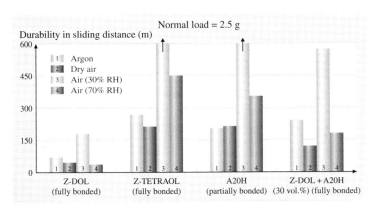

Fig. 40.31 Durability comparison for Z-DOL, Z-TETRAOL, A20H, and Z-DOL + A20H in argon and air with different humidity levels. A single-crystal Si(100) ball (1 mm in diameter) was used as a slider on the films (after [40.62])

that of Z-TETRAOL except in ambient air; both are comparable. Tests on Z-TETRAOL and A20H were terminated at 600 m. Z-DOL + A20H shows higher durability than Z-DOL.

40.4 Nanotribological and Electrical Studies of Ionic Liquid Films

An ionic liquid (IL) is a synthetic salt with a melting point < 100 °C. A room-temperature ionic liquid is a synthetic molten salt with melting point at or below room temperature. One or both of the ions are organic species. At least one ion has a delocalized charge such that the formation of a stable crystal lattice is prevented, and the ions are held together by strong electrostatic forces. As a result of the poor coordination of the ions, these compounds are liquid below 100 °C or even at room temperature [40.20, 21].

The number of combinations of anions and cations that can be used to produce ionic liquids is in the range of one million. Typical cations include imidazolium, pyridinium, ammonium, phosphonium, and sulfonium, as shown in Table 40.4, where "R" stands for an organic group. Typical anions are tetrafluoroborate (BF_4^-), hexafluorophosphate (PF_6^-), bis(trifluorosulfonyl) imide [$(CF_3SO_2)_2N$, *triflamide*], and toluene-4-sulfonate ($C_7H_7O_3S$, *tosylate*) [40.69]. Dependent upon the substrate wettability and other functional requirements, a set of cations and anions can be combined. The ionic liquids were initially developed for use as electrolytes in batteries and for electrodeposition. Recent applications have applied these compounds

Table 40.4 Typical cations

Cation	Structure
Imidazolium	R–N⌐⌐N–R (imidazolium ring, +)
Pyridinium	pyridinium ring N–R, +
Ammonium	NR_4^+
Phosphonium	PR_4^+
Sulfonium	SR_3^+

as environmentally friendly solvents for chemical synthesis (*green chemistry*), where these liquids are used as substitutes for conventional organic solvents.

Ionic liquids are considered as potential lubricants. Their strong electrostatic bonding, compared with covalently bonded fluids, leads to very desirable lubrication properties. They also possess desirable properties such as negligible volatility, nonflammability, high thermal stability or high decomposition temperature, efficient heat transfer properties, low melting point, as well as compatibility with lubricant additives. Unlike conventional lubricants that are electrically insulating, ionic liquids can minimize the contact resistance between sliding surfaces because they are conducting, which is needed for various electrical applications [40.23–25]. These liquids can also be used to mitigate arcing, which is a cause of electrical breakdown in sliding electrical contacts. In addition, ILs have high thermal conductivity, which helps to dissipate heat during sliding. The use of ionic liquids instead of hydrocarbon-based oils (such as highly reformed mineral oils) has the potential to dramatically reduce air emissions. Perfluoropolyethers (PFPEs) are used in magnetic rigid disk and vacuum grease applications due to their high thermal stability and extremely low vapor pressure. However, from the commercial standpoint, ionic liquids are cheaper than PFPEs by a factor of two or so, providing the motivation for comparing the tribological properties of the

Table 40.3 Physical, thermal, and electrical properties of BMIM-PF_6 and Z-TETRAOL

	1-Butyl-3-methylimidazolium hexafluorophosphate (BMIM-PF_6)	Z-TETRAOL
Cation	$C_8H_{15}N_2^+$	–
Anion	PF_6^-	–
Molecular weight (g/mol)	284[a]	2300[b]
$T_{melting}$ (°C)	10[c]	–
T_{decomp} (°C)	300[c]	≈ 320[b]
Density (g/cm^3)	1.37[a]	1.75[b]
Kinematic viscosity (mm^2/s)	281[a] (20 °C)	2000[b] (20 °C)
	78.7[d] (40 °C)	
Pour point (°C)	< -50[e]	-67[b]
Specific heat (J/(g K))	1.44[f] (25 °C)	≈ 0.20[b] (50 °C)
Thermal conductivity at 25 °C (W/(m K))	0.15[g]	≈ 0.09[b]
Dielectric strength at 25 °C (kV/mm)	–	≈ 30[b]
Volume resistivity (Ω cm)	–	$\approx 10^{13}$[b]
Vapor pressure at 20 °C (Torr)	$< 10^{-9}$	5×10^{-7}[b]
Wettability on Si	Moderate[c]	–
Water contact angle	95°	102°
Miscibility with isopropanol	Total[a]	–
Miscibility with water	–	–

[a] Merck Ionic Liquids Database, Darmstadt (http://ildb.merck.de/ionicliquids/en/startpage.htm)
[b] Z-TETRAOL data sheet, Solvay Solexis Inc., Thorofare
[c] *Kinzig* and *Sutor* [40.69]
[d] *Reich* et al. [40.70]
[e] *Wang* et al. [40.71]
[f] *Kabo* et al. [40.72]
[g] *Frez* et al. [40.73]

former with the latter. ILs are being considered for MEMS/NEMS applications because of their high temperature stability, electrical conductivity, and desirable lubrication properties.

Bhushan et al. [40.20] evaluated ionic liquids with the hexafluorophosphate anion deposited on Si(100) wafers, and these were found to exhibit improved friction and wear properties compared with conventional lubricants. The films were evaluated as untreated, partially bonded (by heating at 150 °C for 30 min after dip-coating), and fully bonded (thermally treated and washed). Ionic liquid containing the octyl sulfate anion has also been developed and is of interest due to its resistance to hydrolysis. Based on experience, anions are observed to affect tribological performance. Table 40.3 lists the physical and thermal properties of a selected ionic liquid and its properties, compared with the PFPE lubricant Z-TETRAOL.

The durability of ionic liquid films on various metal and ceramic substrates has been investigated from the standpoint of film formation (wettability) and film removal (friction and wear), where it was found that certain cations and anions exhibit better wetting, friction reduction, and wear resistance properties [40.20]. In general, ionic liquids exhibit better wettability on noble-metal and ceramic surfaces than on nonnoble-metal surfaces [40.69]. The flat imidazolium cation shows poorer wettability compared with bulkier cations such as ammonium and sulfonium. Among salts with the imidazolium cation, the presence of longer organic side-chains leads to reduction of the coefficient of friction. An anion effect is also observed, where oxygen-rich anions show better substrate wettability and lower wear compared with other imidazolium salts. Based on these findings, the ionic liquids of interest are 1-butyl-3-methylimidazolium hexafluorophosphate (BMIM-PF_6) and 1-butyl-3-methylimidazolium octyl sulfate (BMIM-$OctSO_4$). These were studied by *Bhushan* et al. [40.20], and the former was found to be superior in terms of tribological performance. The chemical structures of BMIM-PF_6 and Z-TETRAOL are shown in Fig. 40.32 for comparison, and a summary of their properties is presented in Table 40.3.

Some dicationic ILs are thermally stable up to 400 °C [40.74]. The adhesion and friction properties of two dicationic IL films on Si(100) sub-

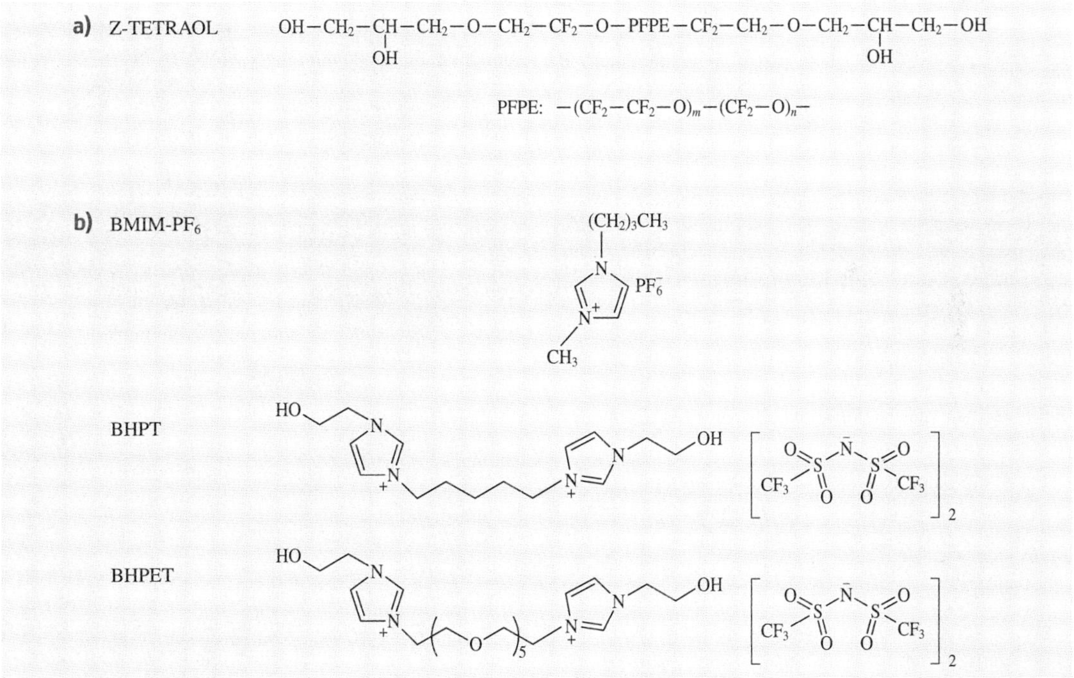

Fig. 40.32 (a) Chemical structures of the Z-TETRAOL molecule, and (b) chemical structures of the BMIM-PF_6, BHPT, and BHPET molecules

strate, based on the imidazolium cation and the bis(trifluoromethanesulfonyl)imide (or *triflamide*) anion, were studied by *Palacio* and *Bhushan* [40.22] and compared with the monocationic IL 1-butyl-3-methyl-1H-imidazolium hexafluorophosphate (BMIM-PF$_6$). AFM experiments were also performed under various humidity and temperature conditions in order to investigate the effect of the environment on the nanolubrication properties of these ILs. Microscale friction and wear experiments using the ball-on-flat tribometer and wear at ultralow loads using an AFM were carried out. Fourier-transform infrared (FTIR) spectroscopy and x-ray photoelectron spectroscopy (XPS) were used to determine the chemical species that affect intermolecular bonding and also to elucidate the effect of the environment on the IL film surface in the case of FTIR data.

The dicationic ionic liquids used in this study were 1,1'-(pentane-1,5-diyl)bis(3-hydroxyethyl-1H-imidazolium-1-yl) di[bis(trifluoromethanesulfonyl)imide] (abbreviated as BHPT) and 1,1'-(3,6,9,12,15-pentaoxapentadecane-1,15-diyl)bis(3-hydroxyethyl-1H-imidazolium-1-yl) di[bis(trifluoromethanesulfonyl)imide] (abbreviated as BHPET) [40.75] deposited on Si(100) wafers. The common name *triflamide* will be used here when referring to the anion. The properties of these dicationic ILs were compared with those of the conventional monocationic ionic liquid 1-butyl-3-methyl-1H-imidazolium hexafluorophosphate, abbreviated as BMIM-PF$_6$ (Merck, Germany). Their chemical structures are shown in Fig. 40.32b. These compounds have been applied on single-crystal Si(100) (phosphorus doped) with a native oxide layer on the surface using the dip-coating technique. The method and the apparatus used have been described earlier. The films were heat-treated at 150 °C for 30 min after dip-coating, such that they were partially bonded [40.20].

40.4.1 Monocationic Liquid Films

In this section, nanotribological data on BMIM-PF$_6$ and Z-TETRAOL are presented [40.20, 21, 25, 46].

Nanotribological Studies

In Fig. 40.33a, the surface height images for the untreated sample (air dried) are compared with the two chemically bonded samples (partially bonded and fully bonded). Aggregates of varying sizes are observed on the untreated Z-TETRAOL and on the ionic liquid coating. These aggregates could have formed initially during preparation of the dilute solution. For Z-TETRAOL, the long PFPE chains can orient in various configurations (such as coils), leading to aggregate formation [40.20]. In ionic liquids containing the 1-butyl-3-methylimidazolium cation, it is believed that aggregate formation in solvents that are less polar compared with water aids in minimizing the charge density (charge delocalization) within the ions. These aggregates are subsequently deposited on the silicon surface during the dip-coating procedure. It is observed that the untreated lubricant surface has more prominent aggregates compared with the two chemically bonded samples, implying that the heat treatment promotes bonding to the Si substrate. Without the chemical bonding procedure, the lubricant molecules are less likely to attach to the substrate and would tend to attract each

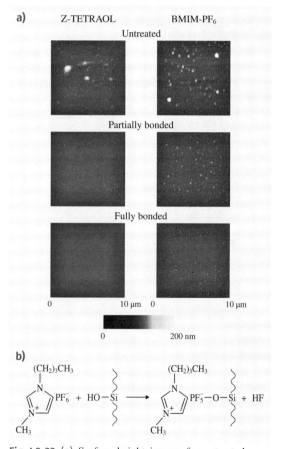

Fig. 40.33 (a) Surface height images for untreated, partially bonded, and fully bonded films of BMIM-PF$_6$ on silicon substrate, and (b) schematic for the attachment of BMIM-PF$_6$ to the silicon substrate (after [40.20])

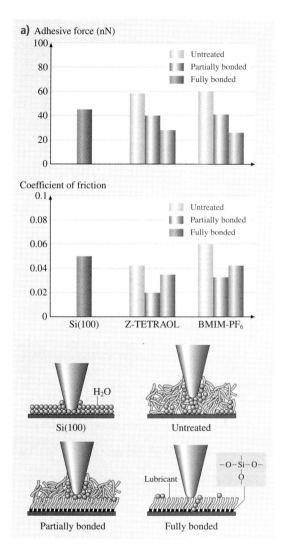
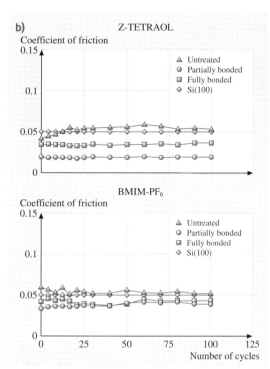

Fig. 40.34 (a) Summary of the adhesive force and coefficient of friction, and (b) durability data after 100 cycles for BMIM-PF$_6$ at room temperature (22 °C) and ambient air (45–55% RH). Data for uncoated Si and Z-TETRAOL are shown for comparison. Schematics in (a) show the effect of chemical bonding treatment and meniscus formation between the AFM tip and sample surface on the adhesive and friction forces (after [40.20])

other instead, such that dewetting is more likely. The immobilization of the ionic liquid, which is promoted by thermal treatment, occurs by reaction of the anion with the hydroxyl groups present on the silicon surface, as shown in Fig. 40.33b for BMIM-PF$_6$.

Figure 40.34a shows a summary of the adhesive force and coefficient of friction measurements on the ionic liquid. Z-TETRAOL and Si(100) data are provided for comparison. The adhesive force has been observed to decrease in the following order: untreated > partially bonded > fully bonded. This mobile fraction on the untreated sample facilitates the formation of a meniscus, which increases the tip–sample adhesion. The adhesive force is highest in the untreated coating since it has the greatest amount of the mobile fraction among the three samples. Conversely, the sample with no mobile lubricant fraction available (fully bonded) has the lowest adhesive force.

A different trend is observed in the coefficient of friction (μ) data. Both the fully bonded and partially bonded samples have lower μ values compared with the uncoated silicon. Friction forces are lower on the latter, implying that the mobile lubricant fraction present in the partially bonded samples facilitates sliding of the tip on the surface. However, μ values for the untreated samples are higher than the data for the heat-treated coatings. Due to the lack of chemical bonding, the interaction of the lubricant with the substrate is weakened

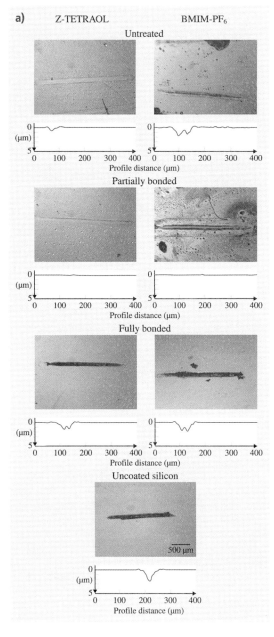
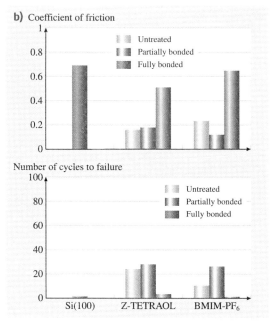

Fig. 40.35 (a) Optical images and height profiles taken after 20 cycles, and (b) summary of the coefficient of friction and number of cycles to failure in ball-on-flat tests on various BMIM-PF$_6$ coatings. Data for uncoated Si and Z-TETRAOL are shown for comparison (after [40.20])

trating the role of meniscus formation in the adhesive and friction forces obtained for the uncoated Si and the untreated, partially bonded, and fully bonded lubricant-coated Si surfaces.

Figure 40.34b shows plots of the coefficient of friction as a function of the number of sliding cycles at 70 nN normal load. Only a small rise in the coefficient of friction was observed for both Z-TETRAOL and BMIM-PF$_6$ surfaces, indicating low surface wear. In the case of untreated Z-TETRAOL, a crossover is observed, where the coefficient of friction increases from its initial value and exceeds the μ of silicon after a certain number of cycles. This is attributed to the transfer of lubricant molecules to the AFM tip and the interaction of the transferred molecules with the lubricant still attached to the Si substrate, which will increase the friction force.

In order to compare friction and wear properties on the nanoscale with that on the microscale, conventional ball-on-flat tribometer experiments were conducted on the same samples. Images and profile traces of the wear scars are shown in Fig. 40.35a. The coefficient of fric-

and dewetting can occur. Water and lubricant molecules are more likely to form a meniscus as the tip approaches the surface. This provides greater resistance to tip sliding, leading to higher coefficient of friction values. The lower portion of Fig. 40.34a shows a schematic illustion and number of cycles to failure are summarized in Fig. 40.35b. Ionic liquid shows enhanced durability compared with both the Z-TETRAOL-coated and the uncoated Si. The nanoscale data presented in Fig. 40.34 can be compared with the μ values obtained from the ball-on-flat measurements (Fig. 40.35b). The μ values of the untreated lubricant samples obtained by using AFM are lower than those obtained from the ball-on-flat tests. This is attributed to the difference in the length scales of the test techniques. An AFM tip simulates a single-asperity contact while the conventional friction test involves the contact of multiple asperities present in the test system [40.1, 2]. With regards to wear, the interface contact of the AFM and ball-on-flat techniques are different from each other. In an AFM, the contact stress is very high, such that material can be displaced more easily. For the ball-on-flat test, the ball exerts a lower pressure on the surface, and the coating is in a confined geometry. As a consequence, displacement of the coating is not as easy as in AFM, leading to enhanced wear resistance.

Wear Detection by Surface Potential and Electrical Resistance Measurements

Figure 40.36a shows a summary of wear tests and corresponding surface potential measurements on the ionic liquid. A bar plot summarizing the average surface potential change on the tested area is shown in Fig. 40.36b. In all cases, a smaller amount of debris was generated compared with the uncoated silicon surface, indicating that the ionic liquid provides wear protection. In general, the samples containing the mobile lubricant fraction (i.e., untreated and partially bonded surfaces) exhibit a lower surface potential change compared with the fully bonded sample, which only has immobile lubricant molecules. This is attributed to lubricant replenishment by the mobile fractions, which can occur in the untreated and partially bonded samples [40.45, 62]. From the bar plot in Fig. 40.36b, it is also observed that the change in surface potential is generally lower in the ionic liquid coating compared with the Z-TETRAOL coating and the uncoated silicon. This indicates that any built-up surface charges arising from the wear test were immediately dissipated onto the conducting ionic lubricant coating surface. In the case of Z-TETRAOL and the uncoated silicon, the charges remained trapped in the test area, since both of these materials are insulators. Based on these findings, a considerable surface potential change will be observed on the wear region when:

1. The lubricant has been fully removed from the substrate;
2. The native SiO_2 layer has been abraded from the surface;
3. Wear has caused subsurface structural changes.
4. Charges build up, as they are unable to dissipate into the surrounding material.

Contact resistance images for the surfaces subjected to the wear tests are presented in Fig. 40.37a. The average change in the contact resistance of the wear region relative to the untested area is summarized in Fig. 40.37b. The fully bonded $BMIM-PF_6$ has an appreciable contact resistance increase in the wear region. Since silicon is a semiconductor, it has much higher resistance compared with the surrounding ionic liquid. The resistance increase in the worn area implies that the substrate is exposed after the wear test. Partially bonded films did not get worn off the substrate by the test, as evidenced by the lack of contact resistance change in the tested area. The untreated Z-TETRAOL exhibited an observable resistance change, while $BMIM-PF_6$ did not. This can be correlated to the durability data in Fig. 40.34b, where the untreated Z-TETRAOL sample exhibited an increase in the friction force with time due to the transfer of lubricant molecules to the tip. Easier lubricant removal means that the diamond tip (in the case of Fig. 40.37) can cause substrate wear much sooner, leading to the observed resistance increase in the tested area. However, the resistance image does not provide a clear contrast between Z-TETRAOL and the newly exposed substrate since both materials have high resistance values.

Microscale contact resistance obtained from ball-on-flat tribometer testing is shown in Fig. 40.38, along with the corresponding coefficient of friction data. For the ionic liquid film, the initial resistance is slightly lower than that of uncoated silicon, confirming their conductive nature. For the Z-TETRAOL samples, the contact resistance is of about the same magnitude as the uncoated silicon. However, for the conducting ionic liquid, an increase in resistance corresponds to an increase in the coefficient of friction, indicating wear of the lubricant and exposure of the silicon substrate, similar to observations on the nanoscale. These results are consistent with the adhesion, friction, and surface potential results with regards to wear detection and wear protection due to the mobile and immobile lubricant fractions.

The durability data and trends for the ionic liquid obtained by using a steel ball are inferior to the re-

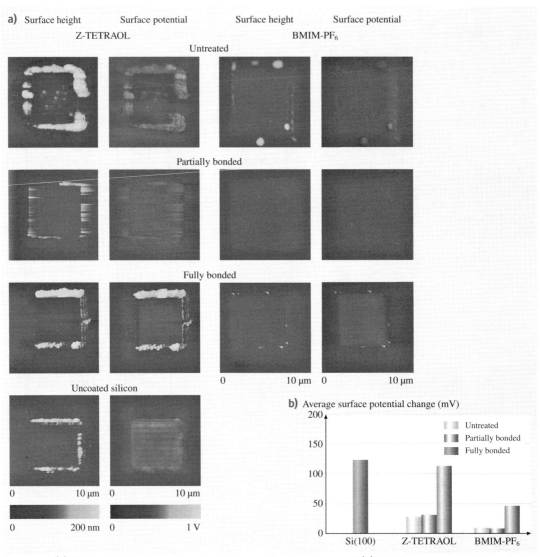

Fig. 40.36 (a) Surface height and surface potential maps after wear tests, and (b) bar chart showing surface potential change for various BMIM-PF$_6$ coatings. Data for uncoated Si and Z-TETRAOL are shown for comparison (after [40.20])

sults shown in Fig. 40.35, which were measured by using a sapphire ball. In Fig. 40.38, the partially bonded samples still show the best durability, but in this case, Z-TETRAOL has the highest number of cycles to failure (the opposite trend compared with Fig. 40.35b), as indicated by the point where the jump in the coefficient of friction is observed. This can be accounted for by the wetting properties of ionic liquids on different surfaces. It has been observed that ionic liquids have a tendency to wet nonmetal surfaces (e.g., Si$_3$N$_4$, SiO$_2$, glass) better than conventional metal surfaces (such as 440C, M50, and 52 100 steel) [40.20]. For wear tests with a steel ball, less wettability means less lubricant retention at the interface. This material wetting effect is possibly more significant for the ionic liquid than for Z-TETRAOL, but nonetheless the durability

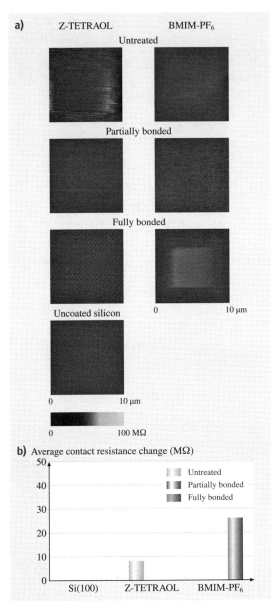

Fig. 40.37 (a) Nanoscale contact resistance images after wear tests, and (b) bar chart showing contact resistance change for various BMIM-PF$_6$ coatings. Data for uncoated Si and Z-TETRAOL are shown for comparison. Cases without the clear wear scar did not exhibit measurable change (after [40.20])

of the partially bonded BMIM-PF$_6$ is still close to its Z-TETRAOL counterpart, such that ILs are still viable lubricants comparable with PFPEs.

40.4.2 Dicationic Ionic Liquid Films

In this section, nanotribological data on BHPT and BHPET are presented. Data on BMIM-PF$_6$ and uncoated Si(100) are also presented for comparison [40.22].

Nanotribological Studies

Figure 40.39a shows a summary of the contact angle, adhesive force, and coefficient of friction (μ) measurements for the coated and uncoated samples performed at ambient temperature and humid conditions (22 °C and 50% RH, respectively). The data shown in the bar plots are averages of three measurements and the error bars represent ±1 σ. BHPT is the least hydrophilic as it has the highest contact angle (81°) among the three IL coatings. This coating also exhibited the greatest reduction in the coefficient of friction relative to the uncoated surface. The high contact angle of BHPT leads to minimal meniscus formation between the tip and surface, leading to a large drop in the nanoscale friction force. In addition, BHPT has a pentyl chain which links the two imidazolium cations. This chain can orient the cation molecules on the substrate, thereby facilitating tip sliding on the film surface. In contrast, the polyether chain of BHPET is susceptible to interactions with water molecules which can promote (instead of minimize) meniscus formation. The BMIM-PF$_6$ film also exhibited a reduction in the adhesive force and coefficient of friction. This results from the combination of mobile and immobile lubricant fractions. Immobilization of this ionic liquid is possible as a result of the thermal treatment, which promotes the reaction between the hexafluorophosphate anion and the hydroxyl groups present on the silicon substrate surface [40.20, 22, 76].

Wear tests were conducted by monitoring the change in the friction force on a 2 μm line for 100 cycles. The data shown in Fig. 40.39b are representative of three measurements made for each sample. The focus of this experiment is the wear of the lubricant film on the substrate. The μ value of the BHPT film changed minimally during the duration of the experiment, which indicates that the film was not being worn after 100 cycles. On the other hand, the BMIM-PF$_6$ and BHPET samples exhibited a gradual increase in μ, which means that these films could be undergoing some

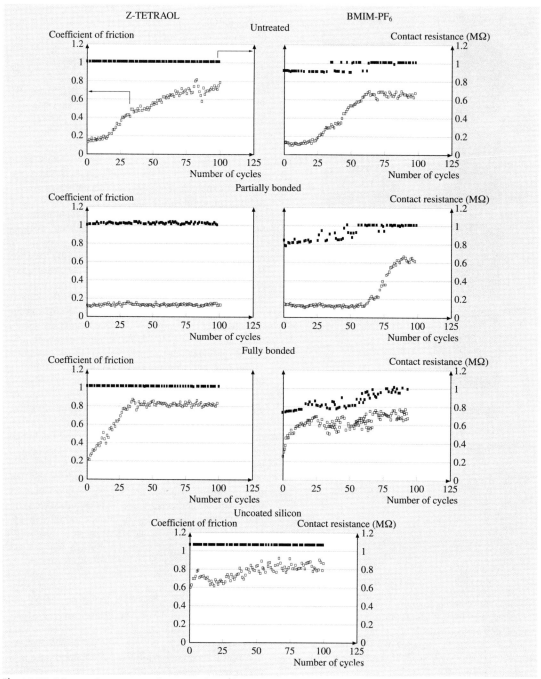

Fig. 40.38 Microscale contact resistance and coefficient of friction after ball-on-flat tests for 100 cycles on various BMIM-PF$_6$ coatings. Data for uncoated Si and Z-TETRAOL are shown for comparison (after [40.20])

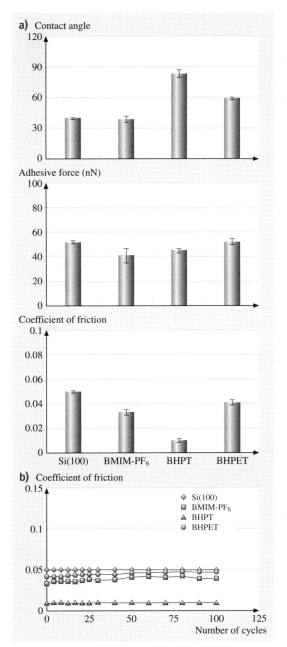

Fig. 40.39 (a) Summary of the contact angle, adhesive force, and coefficient of friction, and (b) durability data after 100 cycles for BMIM-PF$_6$, BHPT, and BHPET coatings at room temperature (22 °C) and ambient air (45–55% RH). Data for uncoated Si are shown for comparison. The *error bars* in (a) represent $\pm 1\sigma$ based on three measurements performed (after [40.22]) ◄

by creating $5 \times 5\,\mu m^2$ wear scars with a diamond tip, where the height and surface potential maps were imaged afterwards. The data shown in the bar plot is the average of three measurements and the error bars represent $\pm 1\sigma$. A change in the surface potential in the wear region is observed when the following occur: the lubricant has been fully removed from the substrate, the native SiO$_2$ layer has been abraded from the surface, wear has caused subsurface structural changes, and charges build up, as they are unable to dissipate into the surrounding material [40.20, 22]. As expected, the uncoated Si exhibited the greatest amount of wear (as evidenced by debris buildup around the edge of the wear test region) and highest increase in the surface potential. The surface potential image for the BHPET film also showed an increase, indicating that the film was worn out after the test. This was not seen on tests with the BMIM-PF$_6$ and BHPT samples. The surface potential change could be absent in the test area if the lubricant was not removed completely, indicating that these two samples have a stronger interaction with the silicon substrate compared with BHPET.

Figure 40.41 presents a summary of the average contact resistance change after the wear test. As in Fig. 40.40, a significant change is observed in the wear

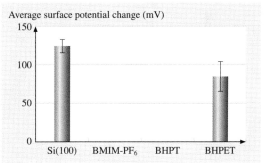

Fig. 40.40 Bar chart showing average surface potential change for BMIM-PF$_6$, BHPT, and BHPET coatings. Data for uncoated Si are shown for comparison. The *error bars* represent $\pm 1\sigma$ based on three measurements performed (after [40.22])

wear and that their interaction with the silicon substrate is weaker compared with that of BHPT.

Figure 40.40 presents a summary of the average contact potential change after the wear tests conducted

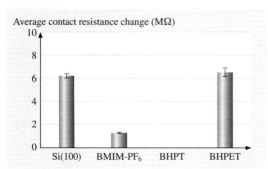

Fig. 40.41 Bar chart showing average contact resistance change for BMIM-PF$_6$, BHPT and BHPET coatings. Data for uncoated Si are shown for comparison. The *error bars* represent $\pm 1\sigma$ based on three measurements performed (after [40.22])

region of the Si and BHPET samples. There is also a small amount of localized contact resistance increase in the BMIM-PF$_6$ sample. These results are consistent with the wear test presented in Fig. 40.39b, where the durability of the films decreases in the order: BHPT > BMIM-PF$_6$ > BHPET.

Microscale Friction and Wear

In order to compare friction and wear properties at the microscale and the nanoscale, conventional ball-on-flat tribometer experiments were conducted on the same samples. The coefficient of friction data are summarized in Fig. 40.42. The data shown in the bar plot are averages of three measurements. All of the lubricated samples are reported to have less wear scars as a result of the ball having to displace the lubricant before damaging the silicon surface. A reduction in the coefficient of friction arising from application of the lubricant film is observed, consistent with the nanoscale adhesion, friction, and wear results.

The μ values of the lubricant samples obtained by using AFM are lower than the μ obtained from the ball-on-flat tests. This is attributed to the difference in the length scales of the test techniques. An AFM tip simulates a single-asperity contact while the conventional friction test involves the contact of multiple asperities present in the test system. With regards to wear, the interface contact of the AFM and ball-on-flat techniques are different from each other such that one cannot expect both tests to show the same trend. On an AFM, the tip stress is very high such that material can be displaced more easily. For a ball-on-flat test, the tip exerts a lower pressure on the surface and the coating is in a confined geometry.

Relative Humidity and Temperature Effect Measurements

The influence of relative humidity on adhesion and friction is summarized in Fig. 40.43. In general, the

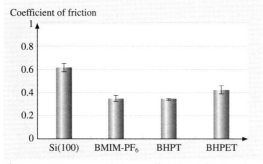

Fig. 40.42 Summary of the coefficient of friction from ball-on-flat tests on uncoated and coated Si samples. The *error bars* represent $\pm 1\sigma$ based on three measurements performed (after [40.22])

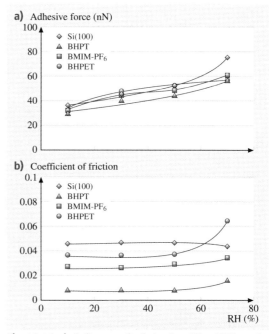

Fig. 40.43a,b Influence of relative humidity (RH) on the adhesive force (**a**) and coefficient of friction (**b**) for unlubricated and lubricated tapes at 22 °C (after [40.22])

adhesive force increases with relative humidity. The condensed water in the humid environment facilitates meniscus formation between the tip and sample and higher adhesive forces. Since Si(100) is hydrophilic, it readily adsorbs water molecules. For the three ionic liquids, an increase in the adhesive force is also due to increased water adsorption. This comes from attractive electrostatic interactions (ion–dipole forces) between the individual ions and water molecules.

Water adsorption affects the coefficient of friction observed as a function of the relative humidity. In Si, the coefficient of friction is uniform at 10–50% RH, then decreases at 70% RH. The adsorbed water at higher humidity can lead to the formation of a continuous water layer separating the tip and sample surface, which can act as a lubricant. Although the adhesive force increases, the reduction in interfacial strength accounts for the slight decrease of the coefficient of friction at the highest range of humidity level examined. However, the presence of more water molecules at higher humidity has the opposite effect on the ionic liquid surfaces, where the coefficient of friction increases with humidity. The attractive ion–dipole forces between the ions and water are amplified at higher humidity because more water molecules are available. A greater attractive force between the tip and the surface leads to greater resistance to sliding and higher coefficient of friction. This is observed for both the monocationic (BMIM-PF$_6$) and the dicationic (BHPT and BHPET) ionic liquid-coated surfaces. For BHPET, polar interactions between water and the oxygen atoms in the polyether (C−O−C) chain are possible and increase the adsorption of water to the surface. This could account for the larger rise in the coefficient of friction in the BHPET sample from 50% to 70% RH, compared with BMIM-PF$_6$ and BHPT.

The effect of temperature on the adhesion and friction properties of the ILs is summarized in Fig. 40.44. The adhesive and friction forces were measured from 22 °C to 125 °C. As shown in Fig. 40.44, the increase in test temperature leads to a decrease in the adhesive force and the coefficient of friction. The decrease in the adhesive force at higher temperatures is observed in all the samples, while the corresponding drop in the coefficient of friction is seen only for BMIM-PF$_6$, BHPET, and the silicon substrate. At higher temperatures, the surface water molecules are desorbed, leading to the decrease in both the adhesive and friction forces. A reduction in the viscosity at higher temperatures can also facilitate the decrease in the friction force [40.26]. In BHPT, the coefficient of friction was not adversely affected as the test temperature was increased. This implies that, under ambient humidity conditions, the BHPT film does not adsorb a large amount of water molecules. Moreover, this also implies that the BHPT surface has weak interactions with surface water molecules, such that the friction force during sliding is not greatly affected.

Fourier-Transform Infrared Spectroscopy

Figure 40.45 presents FTIR spectra obtained for the different ionic-liquid-coated samples, along with the uncoated Si substrate. The observed peaks are labeled with the corresponding chemical bonds. In the case of the Si substrate (with a contact angle of 40°) exposed to water molecules in the ambient, no peaks are observed. C−H stretching vibrations in the coated samples are observed in the 600–800 cm^{-1} range. For BHPET, the strong peak at $\approx 1060\,\mathrm{cm}^{-1}$ corresponds to the C−O−C vibration, which is prominent due to the presence of the polyether chain in its cation. This peak overlaps with the C−O vibration, which is present in BHPT as the terminal primary alcohol (C−OH). In BMIM-PF$_6$, a peak appears in this range due to rock-

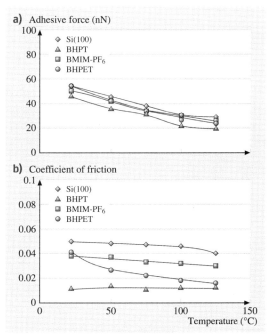

Fig. 40.44a,b Influence of temperature on the adhesive force (**a**) and coefficient of friction (**b**) for uncoated and coated Si samples at 50% RH air (after [40.22])

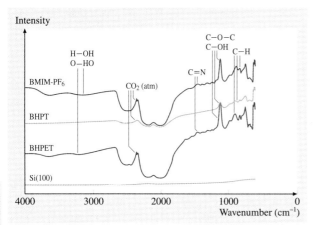

Fig. 40.45 FTIR spectra of uncoated and coated Si samples. The chemical bonds (or species) are listed above the spectra to indicate the possible bonding modes that correspond to the observed peaks (after [40.22])

ing vibrations of the methyl (CH_3) substituent. The peak at $1500-1600\,cm^{-1}$ comes from the C=N vibrations, which is common to all three ionic liquids since they are all based on the imidazolium cation. However, it is not observed in BHPT as the C=N vibrations may be weak. The wide peak at $3600-4000\,cm^{-1}$ corresponds to hydrogen bonding, possibly due to water molecules adsorbed on the surface [40.22, 77, 78]. This is present in BMIM-PF_6 and in BHPET but not in BHPT. This accounts for the much lower contact angle of BMIM-PF_6 and BHPET (39° and 59°, respectively) compared with BHPT (81°). This implies that the surface of BHPT is more hydrophobic compared with that of either BMIM-PF_6 or BHPET, which is consistent with the observed low coefficient of friction for BHPT [40.22].

X-ray Photoelectron Spectroscopy

The XPS survey spectra obtained on the uncoated and coated Si samples are shown in Fig. 40.46a. In the un-

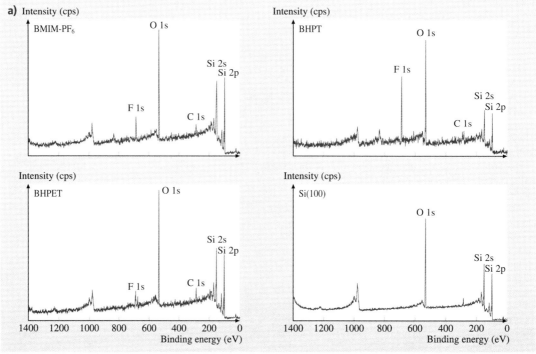

Fig. 40.46 (a) XPS spectra of uncoated and coated Si samples. Survey scan from 0–1400 eV provides the surface elemental composition. (b) High-resolution (deconvoluted) XPS spectra for Si 2p, C 1s, O 1s, and F 1s reveal the different binding environments present on the surface (after [40.22]) ▲ ▶

coated sample, prominent peaks are observed at 99, 151, and 533 eV, corresponding to the binding energies of Si 2p, Si 2s, and O 1s electrons, respectively. Additional peaks are observed on the three coated samples at 285 and 689 eV, which correspond to C 1s and F 1s electrons, respectively. Smaller peaks observed at ≈ 1000 eV on all samples are due to Auger lines (KLL transitions) for oxygen [40.79].

The high-resolution best-fit XPS spectra are shown in Fig. 40.46b for the Si 2p, C 1s, O 1s, and F 1s electrons. The peaks are labeled with the corresponding chemical bonds, which pertain to either the silicon substrate or groups found on the ionic liquid molecule. One noteworthy exception is the presence of peaks at ≈ 292 eV, which confirms the presence of CF_2 on the surface. This indicates the immobilization of the BHPT

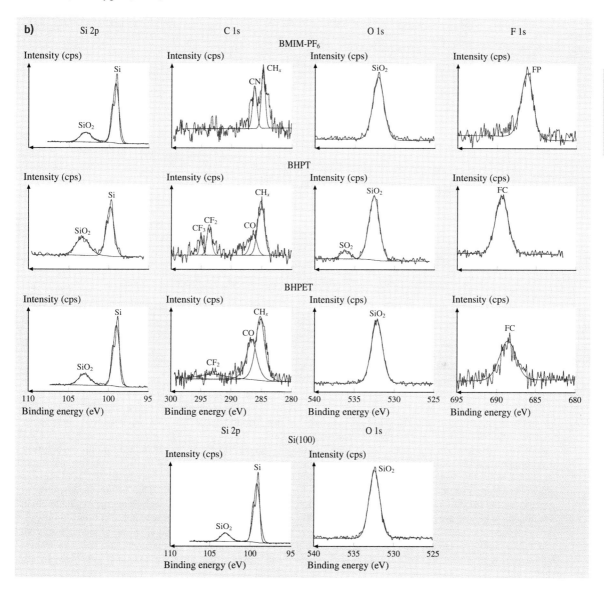

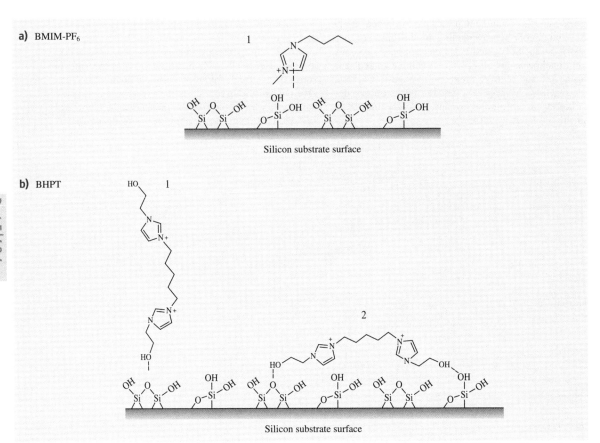

Fig. 40.47a–c Schematic for the attachment of the cations of (**a**) BMIM-PF$_6$, (**b**) BHPT, and (**c**) BHPET to silicon substrate (after [40.22]) ▲ ▶

and BHPET ionic liquids, which occurs by the reaction of the anion with the hydroxyl groups present on the silicon surface [40.22, 23].

Relationship Between IL Structure and Adhesion, Friction, and Wear Properties

Figure 40.47 presents an interpretation of how the IL cations interact with the silicon substrate. For the monocationic BMIM-PF$_6$, only weak interactions between the imidazolium ring and the silicon surface are expected. For the dicationic ionic liquids, multiple cation attachment schemes are possible. In BHPT, the hydroxyl groups attached to the imidazolium cation at the ends of the chain provide a means for strong H-bonding interactions with active sites on the silicon surface. As shown in Fig. 40.47, either one (case 1) or two (case 2) hydroxyl groups can create this bond. The second case is particularly desirable because, if the two hydroxyl groups are bonded (i.e., not exposed to the surface), they are not available to interact with water molecules in the ambient, leading to a reduction in the adhesion and friction forces, as well as enhanced wear resistance [40.22].

In BHPET, these hydroxyl group attachment schemes are also applicable. However, the additional mechanism of intramolecular hydrogen bonding can also take place (case 3). This is not as desirable as the second case because it depletes the available chain ends with hydroxyl groups which can bond to the silicon surface. The interaction of the lubricant film with the silicon substrate is weakened, and water molecules can displace the lubricant from the substrate. In addition, the polyether chain that links the two cations contains five oxygen atoms in each chain. These oxygen

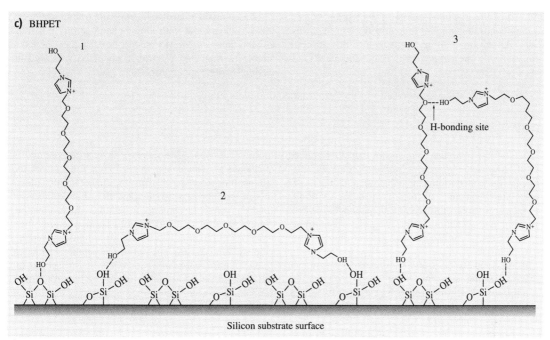

c) BHPET

Fig. 40.47 (continued)

atoms can also form H-bonds with the water molecules in the ambient. This can account for the large difference in the friction properties between BHPET and BHPT. While the application of the BHPT film has lowered the coefficient of friction of the silicon surface, the BHPET film did not have a similar effect due to the molecular interactions described above. This interpretation of lubricant–substrate interaction is corroborated by the adhesion and friction data at varying humidity and temperatures, as well as the FTIR spectra. Both BMIM-PF$_6$ and BHPET have large peaks in their FTIR spectra that correspond to H-bonds of water molecules, which is consistent with the observed sensitivity of their adhesion and friction to the change in humidity and temperature. Meanwhile, BHPT does not have the aforementioned peak, and its adhesion and friction properties appear to be less sensitive to water molecules compared with the other two ionic liquids [40.22].

Figure 40.48 shows a schematic of the interaction of the anions with the silicon substrate. As mentioned previously, XPS spectra indicate immobilization of the ionic liquid, which occurs through the reaction of the anion with the hydroxyl groups on the substrate sur-

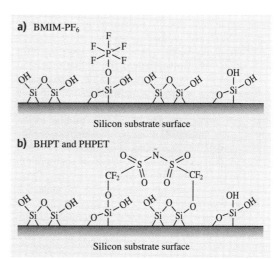

Fig. 40.48a,b Schematic for the attachment of the anions of (**a**) BMIM-PF$_6$, and (**b**) BHPT and BHPET to silicon substrate (after [40.22])

face. In BMIM-PF6, the O attaches to P, while in BHPT and BHPET (which have the same anion), the O bonds to C. Since the reactions involve the creation of new covalent bonds, this implies that the anions of the investigated ILs are more strongly attached to the substrate compared with the cations, which are chemically adsorbed [40.22].

Based on the micro- and nanoscale friction and wear measurements, the ionic liquids show strong potential as lubricants for MEMS/NEMS because they have desirable thermal and electrical conductivity as well as desirable tribological properties.

40.5 Conclusions

Nanodeformation studies have shown that fully bonded Z-DOL lubricants behave as soft polymer coatings, while the unbonded lubricants behave liquid-like. AFM studies have shown that the physisorbed nonpolar molecules on a solid surface have an extended, flat conformation. The spreading property of PFPE is strongly dependent on the molecular end groups and the substrate chemistry.

Using solid-like Z-DOL (fully bonded) film can reduce the friction and adhesion of Si(100), whereas using liquid-like Z-15 lubricant shows a negative effect. Si(100) and Z-15 films show an apparent time effect. The friction and adhesion forces increase as a result of the growth of meniscus up to an equilibrium time, after which they remain constant. The use of Z-DOL (fully bonded) film can prevent rest-time effects. During sliding at high velocity, the meniscus is broken and does not have enough time to rebuild, which leads to a decrease of the friction force and adhesive force for Z-15 and Z-DOL (fully bonded). The influence of relative humidity on friction and adhesion is dominated by the amount of adsorbed water molecules. Increasing humidity can either increase friction through increased adhesion by water meniscus or reduce it through an enhanced water-lubricating effect. Increasing temperature leads to desorption of the water layer, decrease of water surface tension, decrease of viscosity, and easier orientation of the Z-DOL (fully bonded) molecules. These changes cause a decrease of the friction force and adhesion at high temperature. During cycling tests, the molecular interaction between Z-15 molecules attached to the tip and the Z-15 molecules on the film surface causes an initial increase of friction. Wear tests show that Z-DOL (fully bonded) can apparently improve the wear resistance of silicon. Partially bonded PFPE film appears to be more durable than fully bonded films. These results suggest that partially/fully bonded films are good lubricants for micro/nanoscale devices operating under various environmental conditions.

The surface potential technique has been shown to be useful in detecting lubricant removal and initiation of substrate wear. The increase in surface potential is attributed to the change in the work function of the silicon after wear and electrostatic charge buildup of debris in the lubricant. Coatings with a mobile lubricant fraction were better able to protect the silicon substrate from wear compared with the fully bonded coating. This enhanced protection is attributed to a lubricant replenishment mechanism. The contact resistance technique provides information complementary to surface potential data in detecting exposure of the substrate after wear and is a promising method for studying conducting lubricants.

Degradation of novel PFPE lubricants – Z-DOL, Z-TETRAOL, A20H, and Z-DOL + A20H (30 vol. %) – was studied. The coefficient of friction of Z-TETRAOL is lower than that of Z-DOL and A20H in high vacuum. Fully bonded Z-TETRAOL exhibited higher durability than fully bonded Z-DOL on Si, and both are more durable than untreated Z-DOL. A20H shows very high durability in high vacuum. Environment influences the friction and durability of PFPE films. Generally, PFPE films are less durable in high vacuum than in other environments because of intimate contact between the surfaces. In argon, the PFPE films show low friction and high durability. Water molecules can act as a lubricant for PFPE films at a moderate humidity level while they can penetrate the PFPE films and cause increased friction at a high humidity level. The durability of fully bonded Z-TETRAOL and partially bonded A20H is higher than that of Z-DOL in tested environments. The mixture of Z-DOL + A20H (30 vol. %) shows low durability in high vacuum but high durability in argon and air at various humidity levels.

The ionic liquid BMIM-PF$_6$ exhibits low adhesion, friction, and wear properties comparable to the PFPE lubricant Z-TETRAOL. Based on the surface height, adhesion, and friction data, chemical bonding treatment facilitates attachment of the ionic liquid to the silicon substrate surface, leading to a more uniform coating and lower adhesion force and coefficient of friction. The partially bonded coatings have the lowest coefficient of friction and longest durability as they possess a desirable combination of lubricant bonded to the substrate as well as a mobile fraction which facilitates sliding. In micro- and nanoscale experiments, the ionic liquid exhibits durability comparable to that of the lubricant Z-TETRAOL, which has high thermal stability and extremely low vapor pressure. The low postwear surface potential change observed on silicon coated with ionic liquid is indicative of enhanced charge dissipation compared with Z-TETRAOL-coated and uncoated surfaces, which are poor conductors. Contact resistance data is consistent with the surface potential data with regards to identifying the role of the mobile and immobile lubricant fractions in protecting the surface from wear.

The dicationic liquid BHPT exhibits superior nanoscale friction and wear resistance properties. This is attributed to the presence of a pentyl chain and hydroxyl groups on both chain ends, which facilitate molecular orientation as well as bonding interactions with the substrate surface. The other dicationic liquid investigated, BHPET, has less desirable adhesion, friction, and wear properties compared with either BHPT or BMIM-PF_6. Intermolecular hydrogen bonding in BHPET reduces the chain ordering on the substrate surface, which accounts for the observed higher adhesive force and coefficient of friction compared with the other ionic liquids investigated. From the nanoscale wear tests, surface potential, and contact resistance imaging, it was found that the durability of the films decreases in the order: BHPT > BMIM-PF_6 > BHPET. The microscale coefficient of friction is higher than the nanoscale value due to the differences in length scale and configuration of the two test methods.

Nanotribological experiments performed under various humidity and temperature conditions indicate that the adhesive force and coefficient of friction of the ionic liquid films, especially BMIM-PF_6 and BHPET, are highly sensitive to the amount of water molecules present on the surface. This is confirmed by FTIR spectroscopy, where the BMIM-PF_6 and BHPET films show peaks corresponding to hydrogen bonding, mainly resulting from water adsorption and partially from intermolecular interactions. The absence of this peak in BHPT indicates that a much smaller amount of water is adsorbed by this IL, which is consistent with its superior nanotribological properties. The anions are more strongly attached to the substrate compared with the cations. The anions could be covalently bonded, while the cations are chemically adsorbed. XPS analysis confirms the immobilization of the anion onto the silicon substrate.

References

40.1 B. Bhushan: *Principles and Applications of Tribology* (Wiley, New York 1999)
40.2 B. Bhushan: *Introduction to Tribology* (Wiley, New York 2002)
40.3 B. Bhushan: Magnetic recording surfaces. In: *Characterization of Tribological Materials*, ed. by W.A. Glaeser (Butterworth Heinemann, Boston 1993) pp. 116–133
40.4 V.J. Novotny, I. Hussla, J.M. Turlet, M.R. Philpott: Liquid polymer conformation on solid surfaces, J. Chem. Phys. **90**, 5861–5868 (1989)
40.5 V.J. Novotny: Migration of liquid polymers on solid surfaces, J. Chem. Phys. **92**, 3189–3196 (1990)
40.6 C.M. Mate, V.J. Novotny: Molecular conformation and disjoining pressures of polymeric liquid films, J. Chem. Phys. **94**, 8420–8427 (1991)
40.7 C.M. Mate: Application of disjoining and capillary pressure to liquid lubricant films in magnetic recording, J. Appl. Phys. **72**, 3084–3090 (1992)
40.8 G.G. Roberts: *Langmuir–Blodgett Films* (Plenum, New York 1990)
40.9 A. Ulman: *An Introduction to Ultrathin Organic Films* (Academic, Boston 1991)
40.10 B. Bhushan: *Tribology and Mechanics of Magnetic Storage Devices*, 2nd edn. (Springer, New York 1996)
40.11 B. Bhushan: *Handbook of Micro/Nanotribology*, 2nd edn. (CRC, Boca Raton 1999)
40.12 B. Bhushan: Macro- and microtribology of magnetic storage devices. In: *Modern Tribology Handbook, Vol. 2: Materials, Coatings, and Industrial Applications*, ed. by B. Bhushan (CRC, Boca Raton 2001) pp. 1413–1513
40.13 B. Bhushan: *Nanotribology and Nanomechanics – An Introduction*, 2nd edn. (Springer, Berlin, Heidelberg 2008)
40.14 B. Bhushan, Z. Zhao: Macroscale and microscale tribological studies of molecularly thick boundary layers of perfluoropolyether lubricants for magnetic thin-film rigid disks, J. Inf. Storage Proc. Syst. **1**, 1–21 (1999)
40.15 E. Hoque, J.A. DeRose, B. Bhushan, H.J. Mathieu: Self-assembled monolayers on aluminum and copper oxide surfaces: surface and interface characteristics. In: *Applied Scanning Probe Methods IX*, ed. by B. Bhushan, H. Fuchs, M. Tomitori (Springer, Berlin, Heidelberg 2008) pp. 235–281

40.16　B. Bhushan: *Tribology Issues and Opportunities in MEMS* (Kluwer, Dordrecht 1998)

40.17　B. Bhushan: Nanotribology and nanomechanics in nano/biotechnology, Philos. Trans. R. Soc. A **366**, 1499–1537 (2008)

40.18　B. Bhushan, J.N. Israelachvili, U. Landman: Nanotribology: Friction, wear and lubrication at the atomic scale, Nature **374**, 607–616 (1995)

40.19　B. Bhushan: Nanotribology, nanomechanics and nanomaterials characterization, Philos. Trans. R. Soc. A **366**, 1351–1381 (2008)

40.20　B. Bhushan, M. Palacio, B. Kinzig: AFM-based nanotribological and electrical characterization of ultrathin wear-resistant ionic liquid films, J. Colloids Interface Sci. **317**, 275–287 (2008)

40.21　M. Palacio, B. Bhushan: Ultrathin wear-resistant ionic liquid films for novel MEMS/NEMS applications, Adv. Mater. **20**, 1194–1198 (2008)

40.22　M. Palacio, B. Bhushan: Molecularly thick dicationic ionic liquid films for nanolubrication, J. Vac. Sci. Technol. A **27**(4), 986–995 (2009)

40.23　B. Bhushan, K. Kwak: The role of lubricants, scanning velocity and operating environment in adhesion, friction and wear of Pt-Ir coated probes for atomic force microscopy probe-based ferroelectric recording technology, J. Phys. Condens. Matter **20**, 325240 (2008)

40.24　B. Bhushan, K. Kwak, M. Palacio: Nanotribology and nanomechanics of AFM probe based data recording technology, J. Phys. Condens. Matter **20**, 365207 (2008)

40.25　M. Palacio, B. Bhushan: Nanotribological and nanomechanical properties of lubricated PZT thin films for ferroelectric data storage applications, J. Vac. Sci. Technol. A **26**, 768–776 (2008)

40.26　H. Liu, B. Bhushan: Nanotribological characterization of molecularly thick lubricant films for applications to MEMS/NEMS by AFM, Ultramicroscopy **97**, 321–340 (2003)

40.27　J. Ruhe, G. Blackman, V.J. Novotny, T. Clarke, G.B. Street, S. Kuan: Thermal attachment of perfluorinated polymers to solid surfaces, J. Appl. Polym. Sci. **53**, 825–836 (1994)

40.28　J. Ruhe, V. Novotny, T. Clarke, G.B. Street: Ultrathin perfluoropolyether films – influence of anchoring and mobility of polymers on the tribological properties, ASME J. Tribol. **118**, 663–668 (1996)

40.29　V.N. Koinkar, B. Bhushan: Microtribological studies of unlubricated and lubricated surfaces using atomic force/friction force microscopy, J. Vac. Sci. Technol. A **14**, 2378–2391 (1996)

40.30　G.S. Blackman, C.M. Mate, M.R. Philpott: Interaction forces of a sharp tungsten tip with molecular films on silicon surface, Phys. Rev. Lett. **65**, 2270–2273 (1990)

40.31　G.S. Blackman, C.M. Mate, M.R. Philpott: Atomic force microscope studies of lubricant films on solid surfaces, Vacuum **41**, 1283–1286 (1990)

40.32　X. Ma, J. Gui, K.J. Grannen, L.A. Smoliar, B. Marchon, M.S. Jhon, C.L. Bauer: Spreading of PFPE lubricants on carbon surfaces: Effect of hydrogen and nitrogen content, Tribol. Lett. **6**, 9–14 (1999)

40.33　C.A. Kim, H.J. Choi, R.N. Kono, M.S. Jhon: Rheological characterization of perfluoropolyether lubricant, Polym. Prepr. Am. **40**, 647–649 (1999)

40.34　M. Ruths, S. Granick: Rate-dependent adhesion between opposed perfluoropoly(alkylether) layers: Dependence on chain-end functionality and chain length, J. Phys. Chem. B **102**, 6056–6063 (1998)

40.35　U. Jonsson, B. Bhushan: Measurement of rheological properties of ultrathin lubricant films at very high shear rates and near-ambient pressure, J. Appl. Phys. **78**, 3107–3109 (1995)

40.36　C. Hahm, B. Bhushan: High shear rate viscosity measurement of perfluoropolyether lubricants for magnetic thin-film rigid disks, J. Appl. Phys. **81**, 5384–5386 (1997)

40.37　C.M. Mate: Atomic-force-microscope study of polymer lubricants on silicon surface, Phys. Rev. Lett. **68**, 3323–3326 (1992)

40.38　C.M. Mate: Nanotribology of lubricated and unlubricated carbon overcoats on magnetic disks studied by friction force microscopy, Surf. Coat. Technol. **62**, 373–379 (1993)

40.39　S.J. O'Shea, M.E. Welland, T. Rayment: Atomic force microscope study of boundary layer lubrication, Appl. Phys. Lett. **61**, 2240–2242 (1992)

40.40　S.J. O'Shea, M.E. Welland, J.B. Pethica: Atomic force microscopy of local compliance at solid–liquid interface, Chem. Phys. Lett. **223**, 336–340 (1994)

40.41　B. Bhushan, L. Yang, C. Gao, S. Suri, R.A. Miller, B. Marchon: Friction and wear studies of magnetic thin-film rigid disks with glass–ceramic, glass, and aluminum–magnesium substrates, Wear **190**, 44–59 (1995)

40.42　V.N. Koinkar, B. Bhushan: Micro/nanoscale studies of boundary layers of liquid lubricants for magnetic disks, J. Appl. Phys. **79**, 8071–8075 (1996)

40.43　B. Bhushan, S. Sundararajan: Micro/nanoscale friction and wear mechanisms of thin films using atomic force and friction force microscopy, Acta Mater. **46**, 3793–3804 (1998)

40.44　B. Bhushan, C. Dandavate: Thin-film friction and adhesion studies using atomic force microscopy, J. Appl. Phys. **87**, 1201–1210 (2000)

40.45　M. Palacio, B. Bhushan: Surface potential and resistance measurements for detecting wear of chemically-bonded and unbonded molecularly thick perfluoropolyether lubricant films using atomic force microscopy, J. Colloids Interface Sci. **315**, 261–269 (2007)

40.46　M. Palacio, B. Bhushan: Nanotribological properties of novel lubricants for magnetic tapes, Ultramicroscopy **109**(8), 980–990 (2009)

40.47　S. Sundararajan, B. Bhushan: Static friction and surface roughness studies of surface microma-

40.48 B. Bhushan, J. Ruan: Atomic-scale friction measurements using friction force microscopy: Part II – Application to magnetic media, ASME J. Tribol. **116**, 389–396 (1994)

40.49 T. Stifter, O. Marti, B. Bhushan: Theoretical investigation of the distance dependence of capillary and van der Waals forces in scanning probe microscopy, Phys. Rev. B **62**, 13667–13673 (2000)

40.50 J.N. Israelachvili: *Intermolecular and Surface Forces*, 2nd edn. (Academic, London 1992)

40.51 S.K. Chilamakuri, B. Bhushan: A comprehensive kinetic meniscus model for prediction of long-term static friction, J. Appl. Phys. **15**, 4649–4656 (1999)

40.52 H. Ishigaki, I. Kawaguchi, M. Iwasa, Y. Toibana: Friction and wear of hot pressed silicon nitride and other ceramics, ASME J. Tribol. **108**, 514–521 (1986)

40.53 T.E. Fischer: Tribochemistry, Annu. Rev. Mater. Sci. **18**, 303–323 (1988)

40.54 K. Mizuhara, S.M. Hsu: Tribochemical reaction of oxygen and water on silicon surfaces. In: *Wear Particles*, ed. by D. Dowson (Elsevier, New York 1992) pp. 323–328

40.55 S. Danyluk, M. McNallan, D.S. Park: Friction and wear of silicon nitride exposed to moisture at high temperatures. In: *Friction and Wear of Ceramics*, ed. by S. Jahanmir (Dekker, New York 1994) pp. 61–79

40.56 V.A. Muratov, T.E. Fischer: Tribochemical polishing, Annu. Rev. Mater. Sci. **30**, 27–51 (2000)

40.57 H. Yoshizawa, J.N. Israelachvili: Fundamental mechanisms of interfacial friction – Part II: Stick slip friction of spherical and chain molecules, J. Phys. Chem. **97**, 11300–11313 (1993)

40.58 H. Yoshizawa, Y.L. Chen, J.N. Israelachvili: Fundamental mechanisms of interfacial friction – Part I: Relationship between adhesion and friction, J. Phys. Chem. **97**, 4128–4140 (1993)

40.59 K.C. Eapen, S.T. Patton, J.S. Zabinski: Lubrication of microelectromechanical systems (MEMS) using bound and mobile phase of fomblin Z-DOL, Tribol. Lett. **12**, 35–41 (2002)

40.60 D. DeVecchio, B. Bhushan: Use of a nanoscale Kelvin probe for detecting wear precursors, Rev. Sci. Instrum. **69**, 3618–3624 (1998)

40.61 M. Palacio, B. Bhushan: Wear detection of candidate MEMS/NEMS lubricant films using atomic force microscopy-based surface potential measurements, Scr. Mater. **57**, 821–824 (2007)

40.62 Z. Tao, B. Bhushan: Bonding, degradation, and environmental effects on novel perfluoropolyether lubricants, Wear **259**, 1352–1361 (2005)

40.63 B. Bhushan, Z. Tao: Lubrication of advanced metal evaporated tape using novel perfluoropolyether lubricants, Microsyst. Technol. **12**, 579–587 (2006)

40.64 B. Bhushan, M. Cichomski, Z. Tao, N.T. Tran, T. Ethen, C. Merton, R.E. Jewett: Nanotribological characterization and lubricant degradation studies of metal-film magnetic tapes using novel lubricants, ASME J. Tribol. **129**, 621–627 (2007)

40.65 B. Bhushan, J. Ruan: Tribological performance of thin film amorphous carbon overcoats for magnetic recording disks in various environments, Surf. Coat. Technol. **68/69**, 644–650 (1994)

40.66 X. Zhao, B. Bhushan: Comparison studies on degradation mechanisms of perfluoropolyether lubricants and model lubricants, Tribol. Lett. **9**, 187–197 (2000)

40.67 X. Zhao, B. Bhushan: Studies on degradation mechanisms of lubricants for magnetic thin-film rigid disks, J. Eng. Tribol. **215**, 173–188 (2001)

40.68 X. Zhao, B. Bhushan, C. Kajdas: Lubrication studies of head-disk interfaces in a controlled environment – Part 2: Degradation mechanisms of perfluoropolyether lubricants, J. Eng. Tribol. **214**, 547–559 (2000)

40.69 B.J. Kinzig, P. Sutor: *Ionic Liquids: Novel Lubrication for Air and Space*, Phase I, Final Report for AFOSR/NL (Surfaces Research and Applications Inc., Lenexa 2005)

40.70 R.A. Reich, P.A. Stewart, J. Bohaychick, J.A. Urbanski: Base oil properties of ionic liquids, Lubr. Eng. **49**, 16–21 (2003)

40.71 H. Wang, Q. Lu, C. Ye, W. Liu, Z. Cui: Friction and wear behaviors of ionic liquid of alkylimidazolium hexafluorophosphates as lubricants for steel/steel contacts, Wear **256**, 44–48 (2004)

40.72 G.J. Kabo, A.V. Blokhin, Y.U. Paulechka, A.J. Kabo, M.P. Shymanovich, J.W. Magee: Thermodynamic properties of 1-butyl-3-methylimidazolium hexafluorophosphate in the condensed state, J. Chem. Eng. Data **49**, 453–461 (2004)

40.73 C. Frez, G.J. Diebold, C.D. Tran, S. Yu: Determination of thermal diffusivities, thermal conductivities and sound speed of room-temperature ionic liquids by the transient grating technique, J. Chem. Eng. Data **51**, 1250–1255 (2006)

40.74 J.L. Anderson, R. Ding, A. Ellern, D.W. Armstrong: Structure and properties of high stability germinal dicationic ionic liquids, J. Am. Chem. Soc. **127**, 593–604 (2005)

40.75 T. Payagala, J. Huang, Z.S. Breitbach, P.S. Sharma, D.W. Armstrong: Unsymmetrical dicationic ionic liquids: Manipulation of physicochemical properties using specific structural architectures, Chem. Mater. **19**, 5848–5850 (2007)

40.76 M.H. Valkenberg, C. deCastro, W.F. Holderich: Immobilization of ionic liquids on solid supports, Green Chem. **4**, 88–93 (2002)

40.77 R.T. Morrison, R.N. Boyd: *Organic Chemistry*, 6th edn. (Wiley, New York 1992)

40.78 L. Zhang, Q. Zhang, J. Li: Electrochemical behaviors and spectral studies of ionic liquid (1-butyl-3-

methylimidazolium tetrafluoroborate) based on sol–gel electrode, J. Electroanal. Chem. **603**, 243–248 (2007)

40.79 L.C. Feldman, J.W. Mayer: *Fundamentals of Surface and Thin Film Analysis* (Prentice Hall, Upper Saddle River 1986)